Starting Out in Statistics

An Introduction for Students of Human Health, Disease, and Psychology

Patricia de Winter

University College London, UK

Peter M. B. Cahusac

Alfaisal University, Kingdom of Saudi Arabia

WILEY Blackwell

This edition first published 2014 © 2014 by John Wiley & Sons, Ltd

Registered office: John Wiley & Sons Ltd, The Atrium, Southern Gate, Chichester, West Sussex, PO19 8SQ, UK

Editorial offices: 9600 Garsington Road, Oxford, OX4 2DQ, UK
The Atrium, Southern Gate, Chichester, West Sussex, PO19 8SQ, UK
111 River Street, Hoboken, NJ 07030-5774, USA

For details of our global editorial offices, for customer services and for information about how to apply for permission to reuse the copyright material in this book please see our website at www.wiley.com/wiley-blackwell.

The right of the author to be identified as the author of this work has been asserted in accordance with the UK Copyright, Designs and Patents Act 1988.

Library of Congress Cataloging-in-Publication Data

De Winter, Patricia, 1968–
   Starting out in statistics : an introduction for students of human health, disease and psychology / Patricia de Winter and Peter Cahusac.
      pages cm
   Includes bibliographical references and index.
   ISBN 978-1-118-38402-2 (hardback) – ISBN 978-1-118-38401-5 (paper)   1. Medical statistics–Textbooks.   I. Cahusac, Peter, 1957–   II. Title.
   RA409.D43 2014
   610.2′1–dc23
                                                                              2014013803

A catalogue record for this book is available from the British Library.

Wiley also publishes its books in a variety of electronic formats. Some content that appears in print may not be available in electronic books.

Set in 10.5/13pt Times Ten by Aptara Inc., New Delhi, India
Printed and bound in Malaysia by Vivar Printing Sdn Bhd

1   2014

To Glenn, who taught me Statistics
Patricia de Winter

Dedicated to the College of Medicine,
Alfaisal University, Riyadh
Peter M. B. Cahusac

Contents

Introduction – What's the Point of Statistics?

Humans, along with other biological creatures, are complicated. The more we discover about our biology: physiology, health, disease, interactions, relationships, behaviour, the more we realise that we know very little about ourselves. As Professor Steve Jones, UCL academic, author and geneticist, once said 'a six year old knows everything, because he knows everything he *needs* to know'. Young children have relatively simple needs and limited awareness of the complexity of life. As we age we realise that the more we learn, the less we know, because we learn to appreciate how much is as yet undiscovered. The sequencing of the human genome at the beginning of this millennium was famously heralded as 'Without a doubt the most important, most wondrous map ever produced by mankind' by the then US President, Bill Clinton. Now we are starting to understand that there are whole new levels of complexity that control the events encoded in the four bases that constitute our DNA, from our behaviour to our susceptibility to disease. Sequencing of the genome has complicated our view of ourselves, not simplified it.

Statistics is not simply number-crunching; it is a key to help us decipher the data we collect. In this new age of information and increased computing power, in which huge data sets are generated, the demand for Statistics is greater, not diminished. Ronald Aylmer Fisher, one of the founding fathers of Statistics, defined its uses as threefold: (1) to study populations, (2) to study variation and (3) to reduce complexity (Fisher, 1948). These aims are as applicable today as they were then, and perhaps the third is even more so.

We intend this book to be mostly read from beginning to end rather than simply used as a reference for information about a specific statistical test. With this objective, we will use a conceptual approach to explain statistical tests and although formulae are introduced in some sections, the meaning of the mathematical shorthand is fully explained in plain English. Statistics is a branch of applied mathematics so it is not possible to gain a reasonable depth of understanding without doing some maths; however, the most complicated thing you will be asked to do is to find a square root. Even a basic calculator will do this for you, as will any spreadsheet. Other than this you will not need to do anything more complex than addition, subtraction, multiplication and division. For example, calculating the arithmetic mean of a series of numbers involves only adding them together and dividing by however many numbers you have in the series: the arithmetic mean of 3, 1, 5, 9 is these numbers added together, which equals 18 and this is then divided by 4, which is 4.5. There, that's just addition and division and nothing more. If you can perform the four basic operations of addition, subtraction, multiplication and division and use a calculator or Excel, you can compute any equation in this book. If your maths is a bit rusty, we advise that you refer to the basic maths for stats section.

Learning statistics requires mental effort on the part of the learner. As with any subject, we can facilitate learning, but we cannot make the essential connections in your brain that lead to understanding. Only you can do that. To assist you in this, wherever possible we have tried to use examples that are generally applicable and readily understood by all irrespective of discipline being studied. We are aware, however, that students prefer examples that are pertinent to their own discipline. This book is aimed at students studying human-related sciences, but we anticipate that it may be read by others. As we cannot write a book to suit the interests of every individual or discipline, if you are an ecologist, for example, and do not find the relationship between maternal body mass index and infant birth weight engaging, then substitute these variables for ones that are interesting to you, such as rainfall and butterfly numbers.

Finally, we aim to explain how statistics can allow us to decide whether the effects we observe are simply due to random variation or a real effect of an intervention, or phenomenon that we are testing. Put simply, statistics helps us to see the wood in spite of the trees.

Patricia de Winter and Peter M. B. Cahusac

Reference

Fisher, R.A. (1948) *Statistical Methods for Research Workers*, 10th Edition. Edinburgh: Oliver and Boyd.

Basic Maths for Stats Revision

If your maths is a little rusty, you may find this short revision section helpful. Also explained here are mathematical terms with which you may be less familiar, so it is likely worthwhile perusing this section initially or referring back to it as required when you are reading the book.

Most of the maths in this book requires little more than addition, subtraction, multiplication and division. You will occasionally need to square a number or take a square root, so the first seven rows of Table A are those with which you need to be most familiar. While you may be used to using ÷ to represent division, it is more common to use / in science. Furthermore, multiplication is not usually represented by × to avoid confusion with the letter x, but rather by an asterisk (or sometimes a half high dot ·, but we prefer the asterisk as it's easier to see. The only exception to this is when we have occasionally written numbers in scientific notation, where it is widely accepted to use x as the multiplier symbol. Sometimes the multiplication symbol is implied rather than printed: ab in a formula means multiply the value of a by the value of b. Mathematicians love to use symbols as shorthand because writing things out in words becomes very tedious, although it may be useful for the inexperienced. We have therefore explained in words what we mean when we have used an equation. An equation is a set of mathematical terms separated by an equals sign, meaning that the total number on one side of = must be the same as that on the other.

Arithmetic
When sequence matters

The sequence of addition or multiplication does not alter a result, $2 + 3$ is the same as $3 + 2$ and $2 * 3$ is the same as $3 * 2$.

The sequence of subtraction or division does alter the result, $5 - 1 = 4$ but $1 - 5 = -4$, or $4 / 2 = 2$ but $2 / 4 = 0.5$.

Table A   Basic mathematical or statistical calculations and the commands required to perform them in Microsoft Excel where a and b represent any number, or cells containing those numbers. The Excel commands are not case sensitive

Function	Symbol	Excel command	Comments
Addition	+	= $a + b$	Alternatively use the Σ function to add up many numbers in one operation
Subtraction	−	= $a - b$	
Multiplication	*	= $a * b$	
Division	/	= a / b	
Sum	Σ	= $\Sigma(a{:}b)$	See[a] for meaning of $(a{:}b)$
Square	a^2	= $a\hat{\ }2$	Alternatively you may use = $a * a$
Square root	$\sqrt{\ }$	= sqrt(a)	
Arithmetic mean	$\bar{x}$	= average($a{:}b$)	See[a] for meaning of $(a{:}b)$
Standard deviation	s	= stdev($a{:}b$)	
Standard error of the mean	SEM	= stdev($a{:}b$)/sqrt(n)	Where n = the number of observations and see[a] for meaning of $(a{:}b)$
Geometric mean		= geomean($a{:}b$)	See[a] for meaning of $(a{:}b)$
Logarithm (base 10)	$\log_{10}$	= log10(a)	
Natural logarithm	ln	= ln(a)	The natural log uses base e, which is 2.71828 to 5 decimal places
Logarithm (any base)	$\log_a$	= log(a,[base])	Base 2 is commonly used in some genomics applications
Arcsine	asin	= asin(a)	Sometimes used to transform percentage data
Inverse normal cumulative distribution		= invsnorm(probability)	Returns the inverse of the standard normal cumulative distribution. Use to find z-value for a probability (usually 0.975)

[a]Place the cursor within the brackets and drag down or across to include the range of cells whose content you wish to include in the calculation.

Decimal fractions, proportions and negative numbers

A decimal fraction is a number that is not a whole number and has a value greater than zero, for example, 0.001 or 1.256.

Where numbers are expressed on a scale between 0 and 1 they are called proportions. For example, to convert 2, 8 and 10 to proportions, add them together and divide each by the total to give 0.1, 0.4 and 0.5 respectively:

$$2 + 8 + 10 = 20$$
$$2 / 20 = 0.1$$
$$8 / 20 = 0.4$$
$$10 / 20 = 0.5$$

Proportions can be converted to percentages by multiplying them by 100:

$$2 / 20 = 0.1$$
0.1 is the same as 10% i.e. $0.1 * 100 = 10\%$
2 is 10% of 20

A negative number is a number lower than zero (compare with decimal fractions, which must be greater than zero)

Multiplying or dividing two negative numbers together makes them positive, that is,

$$-2 * -2 = 4$$
$$-10 / -5 = 2$$

Squares and square roots

Squaring a number is the same as multiplying it by itself, for example, 3^2 is the same as $3 * 3$. Squaring comes from the theory of finding the area of a square: a square with sides of 3 units in length has an area $3 * 3$ units, which is 9 square units:

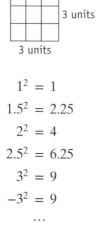

3 units

3 units

$$1^2 = 1$$
$$1.5^2 = 2.25$$
$$2^2 = 4$$
$$2.5^2 = 6.25$$
$$3^2 = 9$$
$$-3^2 = 9$$
$$\ldots$$

Squaring values between 1 and 2 will give answers greater than 1 and lower than 4. Squaring values between 2 and 3 will give answers greater than 4 and lower than 9, etc.

The square sign can also be expressed as 'raised to the power of 2'.

Taking the square root is the opposite of squaring. The square root of a number is the value that must be raised to the power of 2 or squared to give that number, for example, 3 raised to the power of 2 is 9, so 3 is the square root of 9. It is like asking, 'what is the length of the sides of a square that has an area of 9 square units'? The length of each side (i.e. square root) is 3 units:

$$\sqrt{4} = 2$$
$$\sqrt{6.25} = 2.5$$
$$\sqrt{9} = 3$$

Algebra
Rules of algebra

There is a hierarchy for performing calculations within an equation – certain things must always be done before others. For example, terms within brackets confer precedence and so should be worked out first:

$(3 + 5) * 2$ means that 3 must be added to 5 before multiplying the result by 2.

Multiplication or division takes precedence over addition or subtraction irrespective of the order in which the expression is written, so for $3 + 5 * 2$, five and two are multiplied together first and then added to 3, to give 12. If you intend that $3 + 5$ must be added together before multiplying by 2, then the addition must be enclosed in brackets to give it precedence. This would give the answer 16.

Terms in involving both addition and subtraction are performed in the order in which they are written, that is, working from left to right, as neither operation has precedence over the other. Examples are $4 + 2 - 3 = 3$ or $7 - 4 + 2 = 5$. Precedence may be conferred to any part of such a calculation by enclosing it within brackets.

Terms involving both multiplication and division are also performed in the order in which they are written, that is, working from left to right, as they have equal precedence. Examples are $3 * 4 / 6 = 2$ or $3 / 4 * 6 = 4.5$. Precedence may be conferred to any part of such a calculation by enclosing it within brackets.

Squaring takes precedence over addition, subtraction, multiplication or division so in the expression $3 * 5^2$, five must first be squared and then multiplied by three to give the answer 75. If you want the square of $3 * 5$, that is, the square of 15 then the multiplication term is given precedence by enclosing it in brackets: $(3 * 5)^2$, which gives the answer 225.

Similarly, taking a square root of something has precedence over addition, subtraction, multiplication or division, so the expression $\sqrt{2} + 7$ means take the square root of 2 then add it to 7. If you want the square root of $2 + 7$, that is, the square root of 9, then the addition term is given precedence by enclosing it in brackets: $\sqrt{(2 + 7)}$.

When an expression is applicable generally and is not restricted to a specific value, a numerical value may be represented by a letter. For example, $\sqrt{a^2} = a$ is always true whichever number is substituted for a, that is, $\sqrt{3^2} = 3$ or $\sqrt{105^2} = 105$.

Simplifying numbers
Scientific notation

Scientific notation can be regarded as a mathematical 'shorthand' for writing numbers and is particularly convenient for very large or very small numbers. Here are some numbers written in both in full and in scientific notation:

In full	In scientific notation
0.01	1×10^{-2}
0.1	1×10^{-1}
1	1×10^{0}
10	1×10^{1}
100	1×10^{2}
1000	1×10^{3}
0.021	2.1×10^{-2}
25	2.5×10^{1}
345	3.45×10^{2}
4568	4.568×10^{3}

Note that in scientific notation there is only one number before the decimal place in the multiplication factor that comes before the $\times 10$. Where this factor is 1, it may be omitted, for example 1×10^6 may be simplified to 10^6.

Logarithms

The arithmetic expression $10^3 = 1000$. In words, this is: 'ten raised to the power of three equals 1000'. The logarithm (log) of a number is the power to which ten must be raised to obtain that number. So or $\log_{10} 1000 = 3$ or in words, the log of 1000 in base 10 is 3. If no base is given as a subscript we assume that the base is 10, so this expression may be shortened to $\log 1000 = 3$. Here, the number 1000 is called the antilog and 3 is its log. Here are some more arithmetic expressions and their log equivalents.

Arithmetic	Logarithmic
$10^0 = 1$	$\log 1 = 0$
$10^1 = 10$	$\log 10 = 1$
$10^2 = 100$	$\log 100 = 2$
$10^4 = 10,000$	$\log 10,000 = 4$

The log of a number greater than 1 and lower than 10 will have a log between 0 and 1. The log of a number greater than 10 and lower than 100 will have a log between 1 and 2, etc.

Taking the logs of a series of numbers simply changes the scale of measurement. This is like converting measurements in metres to centimetres, the scale is altered but the relationship between one measurement and another is not.

Centring and standardising data

Centering – the arithmetic mean is subtracted from each observation.
Conversion to z-scores (standardisation) – subtract the arithmetic mean from each observation and then divide each by the standard deviation.

Numerical accuracy

Accuracy

Of course it's nice to be absolutely accurate, in both our recorded measurements and in the calculations done on them. However, that ideal is rarely achieved. If we are measuring human height, for example, we may be accurate to the nearest quarter inch or so. Assuming we have collected the data sufficiently accurately and without bias, then typically these are analysed by a computer program such as Excel, SPSS, Minitab or R. Most programs are extremely accurate, although some can be shown to go awry – typically if the data have unusually large or small numbers. Excel, for example, does its calculations accurate to 15 *significant figures*. Nerds have fun showing similar problems in other database and statistical packages. In general, you won't need to worry about computational inaccuracies.

The general rule is that you use as much accuracy as possible **during** calculations. Compromising accuracy during the calculations can lead to cumulative errors which can substantially affect the final answer. Once the final results are obtained then it is usually necessary to *round* to nearest number of relevant *decimal places*. You will be wondering about the specific meanings of technical terms used above (indicated by italics).

Significant figures means the number of digits excluding the zeros that 'fill in' around the decimal point. For example, 2.31 is accurate to 3 significant figures, so is 0.000231 and 231000. It is possible that the last number really is accurate down at the units level, if it had been rounded down from 231000.3, in which case it would be accurate to 6 significant figures.

Rounding means removing digits before or after the decimal point to approximate a number. For example, 2.31658 could be rounded to three decimal places to 2.317. Rounding should be done to the nearest adjacent value. The number 4.651 would round to 4.7, while the number 4.649 would round to 4.6. If the number were 1.250, expressed to its fullest accuracy, and we want to round this to the nearest one decimal place, do we choose 1.2 or 1.3? When there are many such values that need to be rounded, this could be done randomly or by alternating rounding up then rounding down. With larger numbers such as 231, we could round this to the nearest ten to 230, or nearest hundred to 200, or nearest thousand to 0. In doing calculations you should retain all available digits in intermediate calculations and round only the final results.

By now you understand what *decimal places* means. It is the number of figures retained after the decimal point. Good. Let's say we have some measurements in grams, say 3.41, 2.78, 2.20, which are accurate to two decimal places, then it would be incorrect to write the last number as 2.2 since the 0 on the end indicates its level of precision. It means that the measurements were accurate to 0.01 g, which is 10 mg. If we reported the 2.20 as 2.2 we would be saying that particular measurement was made to an accuracy of only 0.1 g or 100 mg, which would be incorrect.

Summarising results

Now we understand the process of rounding, and that we should do this only once all our calculations are complete. Suppose that in our computer output we have the statistic **18.31478642**. The burning question is: 'How many decimal places are relevant'? It depends. It depends on what that number represents. If it represents a statistical test statistic such as z, F, t or χ^2 (Chapters 5, 6, 8), then two (not more than three) decimal places are necessary, for example, 18.31. If this number represents the calculation for the proposed number of participants (after a power calculation, Chapter 5) then people are whole numbers, so it should be given as 18. If the number were an arithmetic mean or other sample statistic then it is usually sufficient to give it to two or three extra significant figures from that of the raw data. For example, if blood pressure was measured to the nearest 1 mmHg (e.g. 105, 93, 107) then the mean of the numbers could be given as 101.67 or 101.667. A more statistically consistent method is to give results accurate to a tenth of their standard error. For example, the following integer scores have a mean of 4.583333333333330 (**there** is the 15 significant figure accuracy of Excel!):

$$3 \quad 2 \quad 2 \quad 3 \quad 5 \quad 7 \quad 6 \quad 5 \quad 8 \quad 4 \quad 9 \quad 1$$

If we are to give a statistic accurate to within a tenth of a standard error then we need to decide to how many significant figures to express our standard error. There is no benefit in reporting a standard error to any more accuracy than two significant figures, since any greater accuracy would be negligible relative to the standard error itself. The standard error for the 12 integer scores above was 0.732971673905257, which we can round to 0.73 (two significant figures). One tenth of that is 0.073, which means we could express our mean between one, or at most two, decimal places. For good measure we'll go for slightly greater accuracy and use two decimal places. This means that we would write our summary mean ($\pm$ standard error) as 4.58 ($\pm$ 0.73). Another example: if the mean were 934.678 and the standard error 12.29, we would give our summary as 935 ($\pm$ 12).

Should we need to present very large numbers then they can be given more succinctly as a number multiplied by powers of 10 (see section on scientific notation). For example, 650,000,000 could be stated as 6.5×10^8. Similarly, for very small numbers, such as 0.0000013 could be stated as 1.3×10^{-6}. The exponent in each case represents the number of places the given number is from the decimal place, positive for large numbers and negative for small numbers. Logarithms are an alternative way of representing very large and small numbers (see section titled Logarithms).

Percentages rarely need to be given to more than one decimal place. So 43.6729% should be reported as 43.7%, though 44% is usually good enough. That is unless very small changes in percentages are meaningful, or the percentage itself is very small and precise, for example, 0.934% (the concentration of Argon in the Earth's atmosphere).

Where have the zeros gone?

In this book, we will be using the convention of dropping the leading zero if the statistic or parameter is unable to exceed 1. This is true for probabilities and correlation coefficients, for example. The software package SPSS gives probabilities and correlations in this way. For example, SPSS gives a very small probability as .000, which is confusing because calculated probabilities are never actually zero. This format is to save space. Don't make the mistake of summarizing a result with $p = 0$ or even worse, $p < 0$. What the .000 means is that the probability is less than .0005 (if it were .0006 then SPSS would print .001). To report this probability value you need to write $p < .001$.

Statistical Software Packages

Statistical analysis has dramatically changed over the last 50 years or so. Here is R. A. Fisher using a mechanical calculator to perform an analysis in the 1940s.

Fortunately, with the advent of digital computers calculations became easier, and there are now numerous statistical software packages available. Perhaps the most successful commercial packages are those of Minitab and SPSS. These are available as stand-alone or network versions, and are popular in academic settings. There are also free packages available by download from the internet. Of these, R is perhaps the most popular. This can be downloaded by visiting the main website http://cran.r-project.org/. R provides probably the most extensive statistical procedures of any of the packages (free and commercial). It also has unrivalled graphical capabilities. Both statistical and graphical procedures are continuously being updated and extended. R initially may be difficult to use for the uninitiated, especially since it is a command line rather than menu-driven package. The extra investment in time to learn the basics of R will be repaid by providing you with greater flexibility, insight and skill. There are numerous guides and blogs for beginners, which can be found by a quick search of the internet. The base R package allows one to do most basic statistical and numerical procedures; however, many other procedures, especially advanced ones, require additional special packages to be installed. This inconvenience is a small price to pay for the greater statistical computing power unleashed. Once a special package has been installed then it needs to be referenced by the command *library(**package.name**)* each time you start a new session. R and its packages are continually being upgraded, so it worth checking every now and then for the latest version. There are integrated development environments, or interfaces, which make using R more convenient and streamlined. In particular, RStudio is recommended. Once R has been installed, RStudio can be downloaded (again free), see http://www.rstudio.com.

All the analyses done in this book will have the commands and outputs using these three packages Minitab, SPSS and R available in Appendix B. In addition, the raw data will be available in .csv format. This will allow you to duplicate all of the analyses.

About the Companion Website

This book is accompanied by a companion website:

www.wiley.com/go/deWinter/spatialscale

The website includes:

- Powerpoints of all figures from the book for downloading
- PDFs of all tables from the book for downloading
- Web-exclusive data files (for Chapters 6, 7 and 10) for downloading

1

Introducing Variables, Populations and Samples – 'Variability is the Law of Life'

1.1 Aims

William Osler, a Canadian physician once wrote: 'Variability is the law of life, and as no two faces are the same, so no two bodies are alike, and no two individuals react alike and behave alike under the abnormal conditions which we know as disease'. We could add that neither do individuals behave or react alike in health either, and we could extend this to tissues and cells and indeed any living organism. In short, biological material, whether it is a whole organism or part of one in a cell culture dish, varies. The point of applying statistics to biological data is to try to determine whether this variability is simply inherent, natural variability, or whether it arises as a consequence of what is being tested, the experimental conditions. This is the fundamental aim of using inferential statistics to analyse biological data.

1.2 Biological data vary

Imagine that you are an alien and land on earth. It seems to be quite a pleasant habitable sort of place and you decide it's worth exploring a little further. It doesn't look like your own planet and everything is new and strange to you. Fortunately, your species evolved to breathe oxygen so you can walk about freely and observe the native life. Suddenly a life form appears from behind

Starting Out in Statistics: An Introduction for Students of Human Health, Disease, and Psychology
First Edition. Patricia de Winter and Peter M. B. Cahusac.
© 2014 John Wiley & Sons, Ltd. Published 2014 by John Wiley & Sons, Ltd.
Companion Website: www.wiley.com/go/deWinter/startingstatistics

some immobile living structures, you later learn are called trees, and walks towards you on all fours. It comes up close and sniffs you inquisitively. You have no idea what this creature is, whether it is a particularly large or small specimen, juvenile or mature, or any other information about it at all. You scan it with your Portable Alien information Device (PAiD), which yields no clues – this creature is unknown to your species. You fervently hope that it is a large specimen of its kind because although it seems friendly enough and wags its rear appendage from side to side in an excited manner, you have seen its teeth and suspect that it could make a tasty meal of you if it decided you were an enemy. If larger ones were around, you wouldn't want to be. You are alone on an alien planet with a strange creature in close proximity and no information. Fortunately, your species is well versed in Statistics, so you know that if you gather more information you will be able to make some assumptions about this creature and assess whether it is a threat to you or not. You climb out of its reach high up into a convenient nearby tree and wait.

You currently have a sample size of one. You need to observe more of these creatures. You don't have long to wait. The life form is soon joined by another of a similar size which sniffs it excitedly. Well, two is better than one, but the information you have is still limited. These two could be similarly sized because they are siblings and both juveniles – the parents could be bigger and just around the corner. You decide to stay put. Some time passes and the pair are joined by 30–40 similar creatures making a tremendous noise, all excited and seemingly in anticipation of something you hope is not you for dinner. Your sample size has grown substantially from two to a pretty decent number. They vary only a little in size; no individual is even close to double the size of another. The creatures are quite small relative to your height and you don't think one ten times the size is very likely to turn up to threaten you. This is reassuring, but you are even happier when a creature you do recognise, a human, turns up and is not mauled to death by the beasts, reinforcing your initial judgement. This example introduces some basic and very important statistical concepts:

1. If you observe something only once, or what a statistician would call a sample size of one, you cannot determine whether other examples of the same thing differ greatly, little or not at all, because you cannot make any comparisons. One dog does not make a pack.

2. Living things vary. They may vary a little, such as the small difference in the size of adult hounds, or a lot, like the difference in size between a young puppy and an adult dog.

3. The more observations you have, the more certain you can be that the conclusions you draw are sound and have not just occurred by chance. Observing 30 hounds is better than observing just two.

In this chapter, we will expand on these concepts and explain some statistical jargon for different types of variables, for example, quantitative, qualitative, discrete, continuous, etc., and then progress onto samples and populations. By the end of the chapter you should be able to identify different types of variables, understand that we only ever deal with samples when dealing with data obtained from humans and understand the difference between a statistical and a biological population.

1.3 Variables

Any quantity that can have more than one value is called a variable, for example, eye colour, number of offspring, heart rate and emotional response are all variables. The opposite of this is a constant, a quantity that has a fixed value, such as maximum acceleration, the speed of light in a vacuum. In the example above, our alien observes the variable 'size of unknown four-legged creature'. While there are some constants in biological material, humans are born with one heart, for example, most of the stuff we are made of falls into the category of variable.

Variables can be categorised into different types. Why is this important in Statistics? Well, later on in this book you will learn that the type of statistical test we use depends in part on the type of variable that we have measured, so identifying its type is important. Some tests can be used only with one type of variable.

First, let us divide variables into two broad categories: those that are described by a number and those that are not. In the following list, which of the variables are not described by numbers?

Eye colour

Number of offspring

Heart rate

Fear

You should have had no difficulty in deciding that the variables 'eye colour' and 'fear' are not described by a number; eye pigmentation is described by colours and fear can be described by adjectives on a scale of 'not fearful at all' to 'extremely'. Or even 'absolutely petrified', if you are scared of spiders and the tiniest one ambles innocently across your desk. We call variables that are not described by a number, *qualitative* variables. You may also hear them called *categorical* variables. It is often stated that qualitative variables cannot be organised into a meaningful sequence. If we were to make a list of eye colours it wouldn't matter if we ordered it 'blue, brown, green, grey' or 'green, blue, grey, brown', as long as all the categories of eye colour are present we can write the list in any order we wish and it would make sense. However, for a qualitative variable such as fear, it would be more logical to order the categories from none to extreme or vice versa.

The two remaining variables on the list above can both be described by a number: number of offspring can be 1, 2, 3, etc. and heart rate is the number of beats per minute. These are *quantitative* or *numerical* variables. Numerical variables have a meaningful progressive order in either magnitude (three offspring are more than two, 80 beats per minute are greater than 60 beats per minute) or time (three days of cell culture is longer than one day)

Some examples of qualitative and quantitative variables are reported in Table 1.1.

Table 1.1   Examples of qualitative and quantitative variables

Qualitative (categorical) variables	Quantitative (numerical) variables
Hair or skin colour	Weight
Nationality or ethnicity	Mean arterial blood pressure
Blood type	Age
Month of birth	Core body temperature
Genotype	Number of offspring
Gender	Blood glucose level
Source of information	Gene expression level
Food preference	Fluorescence intensity

1.4   Types of qualitative variables

Let's take a closer look at qualitative variables. These can be sub-divided into further categories: nominal, multiple response and preference.

1.4.1   Nominal variables

The word nominal means 'pertaining to nouns or names', so nominal variables are those whose 'values' are nouns such as brown, married, alive, heterozygous. The first six qualitative variables in Table 1.1 are nominal. Nominal variables cannot have arithmetic (+, −, ∗, /) or logical operations (>, <, ≥, etc.) performed on them, for example, you cannot subtract French from Dutch, or multiply January by May.

1.4.2   Multiple response variables

This type of variable is frequently found in surveys and questionnaires and is one where a respondent can select all answers that apply. It is a special type of nominal variable. For example, a quality of life survey question might ask prostate cancer patients to select which side effects of anti-androgen therapy they find most unpleasant: hot flushes, difficulty passing urine, swelling or enlargement of the breast, breast tenderness, nausea. As not all patients experience all side effects, study participants would be permitted to select all options that apply to them.

1.4.3 Preference variables

Like multiple response variables, preference variables are also a special type of nominal variable. They are used in surveys and questionnaires and consist of a list of statements, which the respondent must rank in either ascending or descending order. A questionnaire given to patients with Parkinson's disease might ask respondents to score aspects of their health from 1 (most important) to 5 (least important), with each score being used only once. Responses from five patients might look something like the data in Table 1.2. Although the sample size here is small (five), the symptoms that most bother respondents are slow movement and disturbed sleep pattern as these symptoms are ranked more highly than the others. Hence, this type of question aims to identify which variable is most or least preferred from a list and in this case might be used to improve or select treatment options.

Table 1.2 Scores given to five preference variables by five patients

| | Score | | | | |
Symptom	Patient 1	Patient 2	Patient 3	Patient 4	Patient 5
Slow movement	1	2	1	3	1
Tremor	4	3	3	4	4
Constipation	3	5	2	2	3
Abdominal distension	5	4	5	5	5
Disturbed sleep pattern	2	1	4	1	2

1.5 Types of quantitative variables
1.5.1 Discrete variables

This type of variable can have only a whole number as a value. A good example is the number of offspring; one can have one, two, three or more children, but not one-and-a-half. Discrete variables can meaningfully be added, subtracted, multiplied and divided. Logical operations may also be applied, for example, for the variable 'number of offspring', 3 > 2 is a logical statement that makes sense, three children are indeed more than two.

1.5.2   Continuous variables

Continuous variables can have fractional values, such as 3.5 or 0.001; however, we need to subdivide this type of variable further into those that are measured on a ratio scale and those that are measured on an interval scale. The difference between these is whether they are scaled to an absolute value of zero (ratio) or not (interval). A simple example is temperature. In science, there are two scales used for measuring temperature: degrees centigrade or Celsius (°C) and Kelvin (K). Chemists and physicists tend to use predominantly Kelvin, but for convenience biologists often use °C because it's the scale that is used in the everyday world – if you asked me what the core body temperature of a mammal is, I would be able to tell you 37°C without even thinking about it, but I would have to look up the conversion factor and perform some maths if you asked me to tell you in Kelvin. So how do these two scales for measuring temperature differ? Well the value of zero on a Kelvin scale is absolute – it is in effect the absence of temperature – and as cold as anything can be. It is the theoretical value at which there is no movement of atoms and molecules and therefore no production of the heat that we measure as temperature. You cannot have negative values for Kelvin. On the Celsius scale the value of zero is simply the freezing point of water, which is not an absolute zero value as many things can be colder – negative temperatures are not uncommon during winter. One degree Celsius is one-hundredth of the difference between the freezing and boiling point of water, as the latter is given the value of 100°C. So °C is a relative scale – everything is quantified by comparison to these two measurements of water temperature. While ratio scales can have the same mathematical operations applied to them as for discrete variables, interval scales cannot; 200°C is not twice as hot as 100°C, but 200 K is twice as hot as 100 K.

1.5.3   Ordinal variables – a moot point

The final type of variable that we will describe is the ordinal variable. This is a variable whose values are ranked by order – hence *ordinal* – of magnitude. For example, the order of birth of offspring: first born, second born, etc., or the abundance scales used in ecology to quantify number of organisms populating an area with typical ranks from abundant to rare. The reason that we have classified ordinal variable separately is that they are treated as a special type of nominal variable by some and as a numerical variable by others, and there are arguments for and against each.

1.6   Samples and populations

We have now established that there are many types of variables and once we have decided upon which variable(s) we intend to study, we need to decide how we are going to go about it. Let us go back to our alien who landed on our planet with no information about what to expect. He encountered a dog and started making some assumptions about it using what a statistician would call a sample size of one. This minimum sample size yields only a very restricted amount of information, so our alien waited patiently until his sample size grew to a large enough number that he could confidently make a reasoned judgement of the likely threat to his person. Note that the alien did not need to observe all domesticated dogs on earth, a very large sample size, to come to this conclusion. And this is in essence the entire point of Statistics; it allows us to draw conclusions from a relatively small sample of observations. Note that in Statistics we do not use the word observation to mean something that we see, but in a broader sense meaning a piece of information that we have collected, such as the value of a variable whether it be a measurement, a count or a category. We use the information obtained from a small sample to estimate the properties of a whole *population* of observations.

What do we mean by a population? This is where it gets a little confusing because the same word is used to mean different things by different people. The confusion likely arises because biologists, and the public in general, use population to mean a distinct group of individuals such as penguins at the South Pole or cancer patients in the United Kingdom. The definition coined by the influential statistician Ronald Aylmer Fisher (1924) is frequently misrepresented. Fisher stated that a population is in statistical terms a theoretical distribution of numbers, which is not restricted in time or biological possibility: '...since no observational record can completely specify a human being, the populations studies are always to some extent abstractions' and 'The idea of a population is to be applied not only to living, or even to material, individuals'.

So the population of domestic dog sizes – let's take height as the measurement – could include the value 10 metres. Biologically speaking a dog (about the height of four rooms stacked on top of each other), is pretty much impossible, and would make a rather intimidating pet into the bargain. But statistically speaking this height of dog is possible, even if extremely improbable. The chances of meeting such a dog are fortunately so infinitesimally small as to be dismissible in practice, but the *possibility* is real. So a statistical population is quite different from a biological one, because it includes values that may never be observed and remain theoretical. We can ask a computer to provide a sample of values that are randomly selected from a statistical population. We define the numerical characteristics of the population and how many values we would like generated and the software will return our randomly sampled values – most statistical software can do this.

Once we have understood the meaning of a statistical population, it is easier to understand why we only ever have a sample of observations when we collect data from humans, even if we had endless resources and unlimited funding. So, if we were to measure the variable, heart rate, in male athletes, the

Table 1.3   Fictional set of test scores out of 200 for 100 students

45	55	68	111	84	86	107	89	72	92
64	71	12	52	163	19	104	76	86	71
83	89	72	84	99	100	68	70	114	98
93	70	79	69	68	73	68	69	86	38
59	59	79	103	121	105	73	70	83	100
105	76	91	107	122	70	69	84	88	68
67	92	118	39	78	24	73	67	54	62
29	78	51	79	64	135	86	73	110	71
82	97	75	77	75	113	76	77	88	66
63	126	34	56	96	52	82	79	127	72

biological population may be male athletes professionally active within a particular country during a specific period, but the statistical population would be all theoretical values of heart rate, whether observed or not, unrestricted by biological possibility or in time.

When we take a sample of observations, we use this to make some judgements or to use the statistical term, *inferences*, about the underlying *population* of measurements. As a general rule, a very small sample size will be less informative, and hence less accurate than a large sample size, so collecting more data is usually a good idea. This makes sense intuitively – to return to the earlier example of the alien, the greater the number of dogs observed, the more reliable the information about dogs became. If your observations are very variable, then a larger sample size will provide a better estimate of the population, because it will capture more of the variability.

This principle can be demonstrated by sampling from a set of 100 observations (Table 1.3). These observations are a set of random numbers but we shall suppose that they are students' test scores out of 200. This is a sample, not a population, of test score numbers. A set of 100 numbers not organised in any particular order is difficult to interpret, so in Table 1.4 the scores have been ranked from lowest to highest to make it easier for you to see how

Table 1.4   Test scores from Table 1.3 ordered by increasing rank in columns from left to right

12	52	66	69	72	77	83	88	99	111
19	54	67	70	73	77	83	89	100	113
24	55	67	70	73	78	84	89	100	114
29	56	68	70	73	78	84	91	103	118
34	59	68	70	73	79	84	92	104	121
38	59	68	71	75	79	86	92	105	122
39	62	68	71	75	79	86	93	105	126
45	63	68	71	76	79	86	96	107	127
51	64	69	72	76	82	86	97	107	135
52	64	69	72	76	82	88	98	110	163

variable is the whole sample of 100 and that the number in the 'middle' of the sample is in the 70's – we won't worry about exactly what its value is at the moment. Now imagine that an external examiner turns up and wants to know about the test scores but has not seen the data in Tables 1.3 or 1.4 or any of the exam scripts. What would happen if you take just a small number of them rather than show the examiner all the scores?

We can try it. We sample from this set of numbers randomly (using a computer to select the numbers). The first number is 62. Does this single score give us *any* useful information about all the other test scores? Well apart from the fact that one student achieved a score of 62, it tells us little else – it's not even in the middle of the data and is in fact quite a low score. We have no idea from one score of 62 how the other students scored because a sample size of one provides no information about the variability of test scores. Now we'll try randomly selecting two scores: 73 and 82. Well now at least we can compare the two, but the information is still very limited, both scores are close to the middle of the data set, but they don't really capture most of the variability as the test scores actually range from much lower than this, 12, to much higher, 163.

Let's try a much larger sample size of 10 randomly selected scores: 68, 67, 93, 70, 52, 67,113, 77, 89, 77. Now we have a better idea of the variability of test scores as this sample, by chance, comprises some low, middle and higher scores. We could also take a good guess at what the middle score might be, somewhere in the 70's. Let's repeat the sampling several times, say eight, each time taking 11 test scores randomly from the total of 100. The results of this exercise are reported in Table 1.5. The middle score (which is a kind of average called the median) for all 100 scores is 77.5, which is the value exactly between the 50th and 51st ranked scores. We can compare the medians for

Table 1.5 Eight random samples (columns A–H), each containing 11 scores (observations) from Table 1.3. The median score for the original sample of 100 scores is 76.5

	A	B	C	D	E	F	G	H
	38	84	98	99	69	100	52	84
	104	68	113	86	55	39	110	79
	105	79	68	19	78	86	12	86
	62	127	68	56	72	93	52	76
	70	67	105	82	127	24	88	64
	52	79	100	45	76	83	29	92
	79	76	86	68	104	67	84	52
	72	104	88	34	98	59	76	92
	83	77	111	52	89	73	84	70
	97	68	54	77	71	163	73	12
	45	72	135	107	62	70	84	70
Sample median	**72**	**77**	**98**	**68**	**76**	**73**	**76**	**76**

each sample of 11 to see how close they are to the actual median for all samples and we can also compare the range of scores to see how closely they mirror the range of the original sample set. Most of the samples cover a reasonable range of scores from low to high. Half the samples, columns B, E, G and H have medians that differ by only up to 1.3 from the actual median for all scores. One sample, C, contains more middle–high than low scores so the median is much higher at 98. This sample overestimates the test scores and is less representative of the original data. The remaining samples, A, D and F, slightly underestimate the actual median. This example illustrates the importance of unbiased sampling. Here, we have randomly selected observations to include in a sample by using a computer program which is truly random, but suppose you had neatly ordered the test papers by rank in a pile with the highest score on top of the pile and the lowest at the bottom. You hurriedly take a handful of papers at the top of the pile and hand these to the examiner. The sample would then be biased as his scores are those of the top students only. Unbiased sampling is part of good experimental design, which is the subject of the next chapter.

1.7 Summary

- Biological data are variable and if we take measurements of any characteristic of a living organism, we can expect them to differ between individuals. A single observation provides only very limited information as there is no indication of how dispersed a measurement is when there is no other information against which to compare it. Repeated observations are therefore necessary in order to obtain an idea of the variability between individuals.

- Variables may be categorised in many different ways, but broadly fall into two categories: those that are qualitative and those that are quantitative.

- Obtaining all the observations of a particular variable is impracticable and generally impossible, so we take a sample of observations, in order to infer the characteristics of the population in general. In most cases, the larger the sample size, the better the estimate of the characteristics of the population. For sampling to be representative of the population, we would need to avoid bias.

Reference

Fisher, R.A. (1924) *Statistical Methods for Research Workers*. Edinburgh: Oliver and Boyd.

2
Study Design and Sampling – 'Design is Everything. Everything!'

2.1 Aims

The title of this chapter is a quotation from American graphic designer Paul Rand. Doing statistics is inextricably bound up with the data's provenance. A lot of people starting out in statistics think that when we have the data all we need to do is analyse it and get the answer. What is not fully appreciated is the importance of where the data came from: how the research study was designed, how the data were collected. Planning the appropriate design of a study can save tears later on, regardless of the quantity of data collected. The method used to select the data (these could be blood samples, groups of mice or human participants) is crucial to avoid bias. This chapter will focus on these and related issues.

2.2 Introduction

Let us say I saw on the internet that tea tree oil might be effective against spots – this was one person's experience described on a blog. I decide to try this and, after 6 weeks, notice a marked improvement – my spots have vanished! Now, can I persuade you that tea tree oil is a really effective treatment for spots? Maybe, maybe not, it could depend on how loquacious the blog writer was. If not, why not? Well, perhaps you think one person's experience is not enough? Actually the self-experimentation movement has been going from strength to strength. Still, if you are to part with good money for a treatment you might

Starting Out in Statistics: An Introduction for Students of Human Health, Disease, and Psychology
First Edition. Patricia de Winter and Peter M. B. Cahusac.
© 2014 John Wiley & Sons, Ltd. Published 2014 by John Wiley & Sons, Ltd.
Companion Website: www.wiley.com/go/deWinter/startingstatistics

want a 'proper' study done.[1] If the remedy works on a group of people, rather than just one person, then there is a good chance that the effectiveness will *generalise* to others including you. A major part of this chapter is about how we can design studies to determine whether the effects we see in a study can best be attributed to the intervention of interest (which may be a medical treatment, activity, herbal potion, dietary component, etc.).

We've decided that we need more than one participant in our study. We decide to test tea tree oil on 10 participants (is that enough people?). As it happens the first 10 people to respond to your advert for participants all come from the same family. So we have one group of 10 participants (all related to each other) and we proceed to topically apply a daily dose of tea tree oil to the face of each participant. At the end of 6 weeks we ask each participant whether their facial spots problem has got better or worse. There is so much wrong with this experiment you're probably pulling your hair out. Let's look at a few of them:

1. The participants are related. Maybe if there is an effect, it only works on this family (genetic profile), as the genetic similarities between them will be greater than those between unrelated individuals

2. There is only one treatment group. It would be better to have two, one for the active treatment, the other for a placebo treatment

3. Subjective assessment by the participants themselves. The participants may be biased, and report fewer or more spots on themselves. Or individuals might vary in the criteria of what constitutes a 'spot'. It would be better to have a single independent, trained and objective assessor, who is not a participant. It would be ideal if participants did not know whether they received the active treatment or not (single blind), and better still if the assessor was also unaware (double blind)

4. No objective criteria. There should be objective criteria for the identification of spots

5. No baseline. We are relying on the participants themselves to know how many spots they had before the treatment started. It would be better to record the number of spots before treatment starts, and then compare with the number after treatment

You may be able to think of more improvements. I've just listed the glaringly obvious problems. Let's look at some designs.

---

[1] Unbeknown to the author at the time of writing, it turns out that tea tree oil has actually been used in herbal medicine and shown to be clinically effective in the treatment of spots! (Pazyar *et al.*, 2013)

2.3 One sample

The simplest design is to use just one group of people. This is known as a *one-sample* design. Let us say that it was known how spotty people are on average (a very precise value of 3.14159 facial spots). This value is known as the population mean. If we collected a sample of participants who had all been using tea tree oil for 6 weeks, we would count the spots on all the faces of these participants and obtain their arithmetic mean. Our analysis would be to see how this mean differs from the population mean. Obviously, this design is severely limiting in the conclusions that can be drawn from it, for some of the reasons mentioned above in the hypothetical study.

2.4 Related samples

If we have our sample of participants we could test them again. So we observe them (count the number of spots) for a baseline period. We then apply tea tree oil and repeat our observations at a later date. Has there been a change? This is a nice design because it uses each person as their own control. It's called *repeated measures* (because the observations are repeated) or more generally, *related samples* or *within participant* designs. But there can be a problem: perhaps the order in which the treatments are given makes a difference. If we imagine an experiment where we wanted to know whether caffeine improved memory, testing memory first after a decaffeinated drink and then again after a caffeinated drink. The problem is that the participants have done that type of memory test already so, even if new items are used in the test, participants may be more comfortable and able to perform better the second time they get it. The apparent improvement in memory will not be due to caffeine but to practice. It can go the other way too; if participants get bored with the task then they would do worse the second time round. The order of the different types of drinks/treatments themselves could also conceivably make a difference too, leading to a *carry-over* effect. For this reason we can try to control these *order effects* by getting half our participants take the non-caffeinated drink first, and the other half take the caffeinated drink first – this is called *counterbalancing*. Participants are randomly assigned to the order, although there should be approximately equal numbers for the two orders. In medical research this is called a *cross-over* design. Figure 2.1 illustrates this design using the facial spots example where the effects of two types of treatment can be compared. An alternative similar design that avoids this problem, is to use participants that are closely matched in relevant ways (e.g. skin type, age, gender), or use identical twins – if you can get them. Using matched participants is quicker than repeated measures because both treatments can be run at the same time. It is possible to extend this design to more than two treatment conditions.

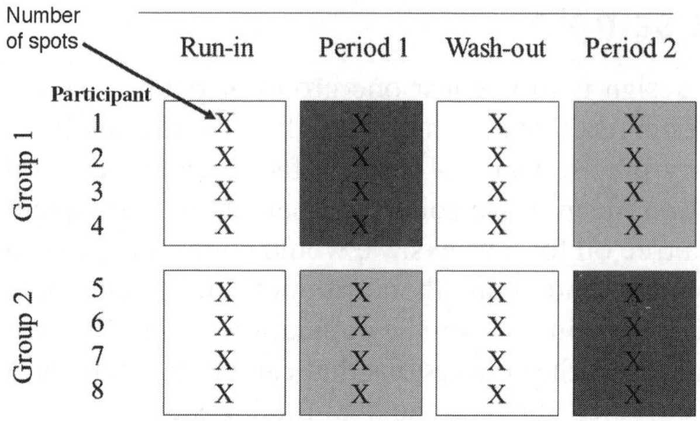

Figure 2.1 In this design two treatments, tea tree oil (dark) and crocus oil (grey) are given at different times. This means that all participants are given both but in a different order (counterbalanced). The Xs represent data values: perhaps the number of facial spots each time they are counted for each person.

Imagine we were interested to see whether tea tree oil worked for particularly spotty people. A sensible thing to do, you might think, would be to find the very spottiest people for our study. We would assess their spottiness before the treatment and then again after 6 weeks of tea tree oil treatment. You'll notice this is a repeated measures design. Now, since spottiness varies over time, those participants we selected to study were likely to be at a peak of spottiness. During the course of the treatment it is likely that their spots will resume their more normal level of spottiness. In other words, there will appear to be a reduction in spots among our spotty recruits – but only because we chose them when they were at the worst of their spottiness. This is known as *regression to the mean* and can be the cause of many invalid research findings. For example, we may be interested in testing whether a new hypotensive drug is effective at reducing blood pressure in patients. We select only those patients who have particularly high blood pressure for our study. Again, over time the average blood pressure for this group will decline anyway (new drug or not), and it will look like an effect of the medication, but it isn't.

2.5 Independent samples

A more complicated study would randomly divide our participants into two groups, one for treatment, the other for control. First, why 'randomly'? If we did not randomly allocate participants to the groups then it is possible that there may be some bias (intentional or unintentional) in who gets allocated to the groups. Perhaps the non-spotty-faced people get allocated to the treatment (tea tree oil) group and the spotty-faced people allocated to the control group. If so, we already have a clear difference even before the treatment begins. Of course the bias in allocation could be more subtle, such as age

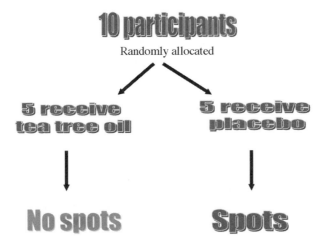

Figure 2.2 Independent samples or parallel group designs. Participants are randomly allocated to receive different treatments. After treatment the effectiveness of the treatment is assessed. In this study all those receiving tea tree oil end up with no spots, while all those receiving placebo have spots. It very much looks like tea tree oil helps reduce spots.

or skin type. We must avoid any subjective bias by the investigator in group allocation. Well, one group receives the tea tree oil, the other group nothing (control). After 6 weeks, the number of spots are compared between the two groups. It may be that any kind of oil, not just tea tree oil helps reduce spots. A better design would use an alternative oil, such as mineral oil, as a *control*. Better still, don't tell the participants which treatment they are receiving so that the control now becomes a *placebo* group (and called *single blind* because they don't know). If the spot assessor also didn't know who had received which treatment, this would then be called *double blind*. Human nature being as it is, it is best to avoid subjective biases by using blind or double-blind procedures. This design is often called *independent samples* design or, in medical settings, *parallel groups*, and is a type of *randomised control trial* (RCT), see Figure 2.2. The general term for this type of design is *between participants* because the treatments vary between different participants (cf. within participants design above). It is often quicker to do than repeated measures as everyone may be tested at the same time, but it requires more participants in order to obtain convincing results – as each person does not act as their own control, or have a matched control, unlike the related samples design.

2.6 Factorial designs

A more elaborate form of parallel groups or independent samples is to have more than one variable under the investigator's control. As above, one variable could be the type of oil used (tea tree vs. placebo). We could introduce

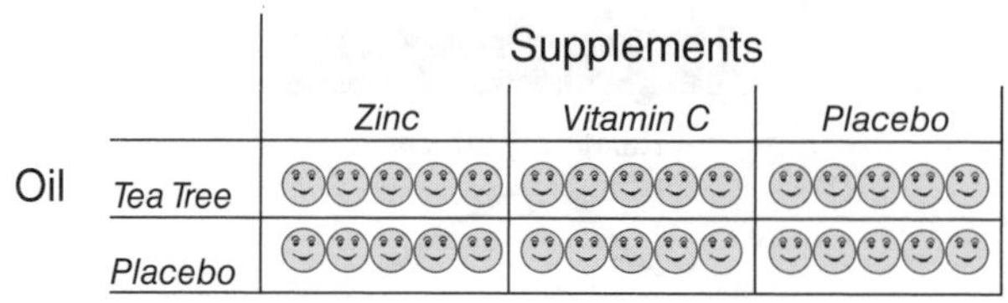

Figure 2.3 Studying two independent variables simultaneously. Here, if we had 30 participants, they would be randomly assigned to each of the six different subgroups (five in each).

another variable, for example, a dietary supplement variable: zinc supplement, vitamin C supplement, and a placebo supplement. Participants would need to be randomly assigned to the type of supplement just like they were to the tea tree oil/placebo treatments. This means we would have two variables in our design, the first variable (oil) with two *levels*, and the second variable (supplements) with three levels, see Figure 2.3. This allows us to see, overall, whether tea tree oil has an effect, but simultaneously whether one or other of the supplements help. This type of study is known as a *factorial* design, because we are examining the effects of more than one variable (factor). The effects seen across each of the different variables are known as *main* effects. So, we are getting two studies for the price of one. Actually we get more, because we can also see whether there might be some *interaction* between tea tree oil and the supplements. For example, perhaps individual treatments on their own don't do very much, but the subgroup which receives both tea tree oil and zinc together might show synergism and produce a large amelioration in spots. If there is an interaction present then we can identify its nature by looking at *simple effects* analyses. A simple effects analysis is when we look at differences between the different levels of one variable at just one level of the other variable. This is done for each level of the second variable, and vice versa, all levels of the second variable are compared at each level of the first variable.

We could also have one of the variables not under the investigator's direct control, such as gender (it is not possible to randomly assign people to gender!). When we do not have direct control in group allocation (as for gender) this makes it a *quasi-experimental* design. Again we could determine the main effect for the oil, the main effect for gender (are there differences between men and women?), as well as the interaction between the variables. In fact we could even include gender as a third variable, giving us three main effects and four different interactions between the three variables (Oil * Supplements, Oil * Gender, Supplements * Gender, Oil * Supplements * Gender).

Finally, it is possible to introduce repeated measures (within participant variables) into such an experimental design. We could, for example, have a cross-over design with both types of oil given to male and female participants at different times (half would receive tea tree oil first, the other half would

receive placebo oil first). This gives us a *mixed* within and between participants design.

2.7 Observational study designs

So far, we have been talking about *experimental* designs. These are useful
designs in order to help determine whether a treatment actually has an effect:
that there is a causal relationship between the tea tree oil and a reduction
in spots. However many studies, especially in epidemiology, involve *observational* studies. This is where we are unable, either for practical or ethical
reasons, to intervene with a treatment. Instead we must be content to merely
observe our participants and see whether what happens to them is correlated
with what we are interested in (e.g. whether the spots get better or worse).
This is a weaker approach because correlation does not mean causation. However, it is often a useful way of doing *exploratory studies* as a prelude to doing
properly controlled experiments to confirm causality between one or more
independent variables and an outcome.

2.7.1 Cross-sectional design

For this design we are looking at a cross-section through a population at a
specific point in time. For example, we could take a representative sample
from the population of interest and determine the prevalence of facial spots.
This means we are taking a snapshot of people at one particular instance in
time. We can also survey participants about diet, age, gender, etc. and see
which of these factors correlate with facial spots. However, we need to be
wary of assigning a causal role to any of these factors as the correlation may be
spurious, for example, it may not be diet that is causing the spots but hormonal
changes in teenagers who tend to have poor diets. Diet would also be known
as a *confounding variable*, or *confound* – for the same reason. Correct multivariate analysis of the data may be able to identify and take into account such
confounds.

2.7.2 Case-control design

Another approach is to identify individuals who have a spots problem and
compare them with individuals with whom they are **matched** for age, gender,
socioeconomic status, etc. The task then becomes to identify a specific difference between these two groups that might explain why one group is spottier
than the other. It might be diet or exercise, exposure to pollutants or any number of factors. This type of study is known as a *case-control* design. If one of
the differences was that the spotty group used more facial remedies for spots,
then this might identify the facial remedies as causes, when in fact they are

attempts by individuals to alleviate their spots! This is similar to the everyday observation that diet drinks must 'make' people fat because most fat people are drinking them. Case-control studies are relatively cheap to do and are popular in epidemiology. However, they are also a rather weak source of evidence since we usually do not know accurately the exposure history of the patients and matched controls. Despite this drawback, this design did provide the first evidence for the link between lung cancer and tobacco use.

2.7.3   Longitudinal studies

There are also studies in which we observe people over long periods of time to see what changes happen. These are known as *longitudinal* studies. In one type of study, participants are chosen because they were born in a particular year or place. Even a whole country's population has been studied – for example, the mental health of the 7.25 million population of Sweden (Crump *et al.*, 2013). The selected group of people are then followed over time, sometimes for many years. We observe which people succumb to a particular disorder and then correlate that with factors that might have precipitated the disorder. We might, for example, determine a factor, say environmental or dietary, which seems to be strongly associated with people developing facial spots. This type of study is known as a *cohort* design, and is often prospective in that we are looking forward in time when gathering the data. A cohort study can also be done on archival data making it retrospective. As described above, we again may have problems attributing the cause to a specific factor, and typically many other factors need to be taken into account (*controlled* for) in order to identify the possible guilty factor(s).

2.7.4   Surveys

Finally, *surveys* are a popular and convenient way of collecting data. These often employ a *questionnaire* with a series of questions. Items in the questionnaire might require a simple Yes/No response, or they may allow graded responses on a scale 1–5 (known as a *Likert* scale), for example, where 1 might be 'strongly disagree' and 5 might be 'strongly agree' to some statement asked – such as 'I am happy in my employment'. With surveys we need to obtain a representative sample from our population of interest – our *sampling frame*. The best way to do this is by *random sampling* of people from that population. Even then it is highly unlikely that all people sampled will agree to respond to your questionnaire, and so you must be very aware of the possible *bias* introduced by non-responders. If the population of interest is distributed over a large area, for example, the whole country, then it would be impractical to travel to administer the questionnaire to all the selected individuals in your random sample. One convenient way is to use *cluster sampling*, where geographical clusters of participants are randomly obtained, and the participants within each cluster are administered the questionnaire. Variations of cluster sampling may also be used in designs described earlier. We may want to ensure that men and women are equally represented in our sample, in which case we would use *stratified* sampling to ensure the number of men and women sampled correspond to their proportions in the sampling frame (population) – see next section for a fuller explanation of stratified.

2.8 Sampling

We have already mentioned sampling above. Whether we select participants for an experimental study or for an observational study, we need to draw conclusions about the population from our sample. To do this our sample must be *representative*, that is, it should contain the same proportion of male and female, young and old and so on, as in the population. One way to try to achieve this is by taking a random sample. The word random has a very specific meaning in statistics, although random is used colloquially nowadays to also mean unusual for example, 'a random woman on the train asked me where I'd bought my coat' or 'the train was late because of a random suspect package'. The word random occurs in several contexts in statistics. In sampling, it means that each item has an equal probability of being selected. This is in the same sense of a lottery winning number being selected at random from many possible numbers. A *random variable* means that the values for that variable have not been selected by the investigator. Observations can be regarded as being subject to random variability or error, meaning the variation is uncontrolled, unsystematic, and cannot be attributed to a specific cause.

Taking a random sample will help ensure the *external validity* of the study – our ability to draw correct conclusions about the population from which our sample was selected. The ideal that we should aspire to, but rarely achieve, is *simple random sampling* in which each item within the sampling frame has an equal chance of being selected. The larger the sample the more accurate it will be. If we are interested in proportions of participants, for example, having a particular view on an issue, then the *margin of error* (95% confidence interval – discussed later in Chapter 5) will not exceed the observed proportion $\pm \sqrt{N}$, where N is the size of the sample. Moreover, within limits, the level of a sample's accuracy will be independent of the size of the population from which it was selected: that is, you do not need to use a larger sample for a larger population. It should be remembered that random sampling will not guarantee that a representative sample will be obtained, but the larger the sample the less chance there would be of obtaining a misleading sample. Sometimes *strata* are identified within a population of interest. Strata, similar to its geological use, represents *homogenous* (similar to each other) subgroups within a population, for example, gender, race, age, socioeconomic status, profession, health status, etc. Strata are identified before a study is carried out. Sampling is said to be stratified if a sample contains predefined proportions of observations in each stratum (subgroup). It helps ensure that the sample is representative by including observations from each stratum (so that important subgroups (e.g. minorities) are not excluded during the random selection of a sample). It is assumed that members within each stratum are selected at random, although in practice selection may not be truly random. If a variable used in stratification, such as gender, is correlated with the measure we are interested in, then the sample stratified according to that variable will provide increased accuracy.

Quota sampling is where one or more subgroups are identified, and specified numbers of observations (e.g. participants) are selected for each subgroup. The sampling is typically non-random, and is often used in market research of customers to enhance business strategy, for example, to survey views about car features from car drivers (subgroups could be according to the drivers' class of vehicle: saloon, estate, sports, sports utility, etc.).

In some areas of social psychology the investigator may be interested in a difficult-to-access ('hidden') population, for example, drug users. In this case *snow-ball* sampling can be used in which an identified participant may provide the investigator with contacts to other participants, who then identify further participants, and so on.

In many experimental studies, participants are typically obtained by recruiting them through poster adverts. They are often university students or members of the public in a restricted geographical area. This type of sample is called a *convenience* or *self-selecting* sample because minimal effort is made to randomly select participants from the population. However, it is important to randomly allocate these participants to the different treatment conditions, and this helps ensure that the study has *internal validity* – meaning that the effects seen can be reasonably attributed to the intervention used.

2.9   Reliability and validity

It goes without saying that when we take measurements these should be done carefully and accurately. The degree to which measurements are consistent and do not contain measurement error is known as *reliability*. Someone else repeating the study should be able to obtain similar results if they take the same degree of care over collecting the data. Speaking of the accuracy of measurements, it is worth mentioning that there will often be some kind of error involved in taking them. These can be small errors of discrepancy between the actual value and what our instrument tells us. There will also be errors of judgement, misreadings and errors of recording the numbers (e.g. typos). Finally, there are individual differences between different objects studied (people, animals, blood specimens, etc.), and statisticians also call this 'error', because it deviates from the true population value. These errors are usually random deviations around the true value that we are interested in, and they are known as *random error*. This source of variability (error) is independent of the variability due to a treatment (see Chapter 6). We can minimise their contribution by taking greater care, restricting the type of participants (e.g. restrict the age range), and by improving the precision of our instruments and equipment. Another type of error is when the measured values tend to be in one direction away from the true value, that is, they are biased. This is known as *systematic error*, and may occur because the investigator (not necessarily intentionally) or the instrument shows consistent bias in one direction. This means that other investigators or other instruments will not replicate the findings. This is potentially a more serious type of error, and can be minimised by the investigator adopting a standardised procedure, and by calibrating our instruments correctly. Systematic bias can also occur due to participant variables like gender or age, and may bias the findings. If this bias is associated with the independent variable, then the bias becomes a *confounding* variable (or *confound*). Their effect should be controlled by random assignment of participants to different treatment conditions, but may still be problematic.

A rather less understood or appreciated concept is *validity*. When we use a tape measure to measure someone's height it is intrinsic to the procedure (manifestly true) that what we are doing is actually measuring height (give or take random errors). However, if I were to accurately measure blood pressure in participants and then claim that I was measuring how psychologically stressed they were, you would rightly question my claim. After all, blood pressure can increase for many reasons other than stress: excitement, anger, anxiety, circulatory disorders, age. Validity is all about the concerns over the usefulness of a measure, clinical test, instrument, etc. – whether it is really measuring what is claimed or intended to be measured. If the measurements lack validity, this will typically bias them, resulting in systematic error referred to earlier. Often the way to measure something as elusive as 'stress', known as a *latent* variable, could be to use more than one instrument or test (e.g. blood pressure, blood cortisol level, questionnaire) in the study to provide a more valid measure or *profile* for the condition we call stress. Even then there may be some dispute among researchers and theoreticians about the real definition of what stress really is.

2.10 Summary

- Study design is an important part of carrying out a research project. The design will dictate how the data are collected and analysed. A poor design might mean that no clear conclusion can be drawn from the study.

- The simplest design consists of one sample of observations. This is fine if we have a specific value for a parameter (say a population mean) with which we can compare our sample's value. Otherwise we have no other treatment to compare it with (like a control group), nor do we have a baseline.

- The related samples design is very useful. We have a comparison with baseline measurements or with matched participants (twins are ideal). This may be extended to more than two treatment conditions. However, order effects may be a problem for repeated measurements on the same individuals. These need to be combated using counterbalancing. We need to be wary about the issue of regression to the mean with this design.

- Independent samples allow us to compare unrelated groups, for example, a treatment group with a placebo group. This has no order effect problem but requires more participants than the related sample design.

- The factorial design allows more than one independent variable. This allows the effects of two or more variables to be observed in a single study. It also allows us to see if there is an interaction between the variables. This design may be extended to a mixed design which includes one or more within participant variables (e.g. related sample measurements).

- Observational studies are an important source of information obtained without intervening or controlling variables. They are useful in observing correlations, but generally do not allow us to determine causality between variables. They can be useful as an exploratory stage as a prelude to a properly controlled experiment.

- A number of different types of observational designs are used:

  Cross-sectional: looking at a population at a particular point in time.

  Case-control: affected individuals are matched with controls and their history compared.

  Longitudinal: these may be prospective (looking ahead) or retrospective (looking into the past).

- Surveys are a popular way of collecting data. Usually a questionnaire is used. The difficulty is obtaining a random sample, either because of non-random selection of participants or non-response by participants.

- Sampling is fundamental for ensuring the external validity of research designs. The key issue is that our sample must be representative of the population we are interested in. This usually means we should select a random sample. Different types of sampling (e.g. cluster, snow-ball, convenience) are used depending on the context and constraints of the study.

- Reliability concerns the accuracy of the measurements. Measurement error should be minimised.

- Validity is the simple, but unappreciated, concept that what we are measuring is actually what we think we are measuring and making claims about.

References

Crump, C., Sundquist, K., Winkleby, M.A. and Sundquist, J. (2013) Mental disorders and vulnerability to homicidal death: Swedish nationwide cohort study. *British Medical Journal*, 346:f557.

Pazyar, N., Yaghoobi, R., Bagherani, N. and Kazerouni, A. (2013) A review of applications of tea tree oil in dermatology. *International Journal of Dermatology*, 52(7):784–790.

3

Probability – 'Probability ... So True in General'

3.1 Aims

The title is a quotation from English historian Edward Gibbon: 'The laws of probability, so true in general, so fallacious in particular'. Probability is the key to our understanding of statistics. In simple terms, probability represents the chance of an event happening. It is often used to express how certain we might be about the truth of some situation or the cause of an outcome (e.g. how certain can we be that this person committed the crime or, that this disease Y is caused by X?). It always represents varying degrees of uncertainty, and all statistics are expressed in terms of uncertainty. Probability is not the easiest concept to understand, but we will need to develop some familiarity with it. Historically, there have been several approaches to statistical testing. They have been associated with different interpretations of probability and differences in how probability is expressed. This chapter represents an introduction to the topic.

3.2 What is probability?

It was a desperate strategy, based on probabilities, but it was all he had left.
Robert Ludlum, The Bourne Supremacy

In statistics, drawing conclusions from data always involves uncertainty. The numbers that we've collected may be unusual in some way (perhaps we were

Starting Out in Statistics: An Introduction for Students of Human Health, Disease, and Psychology
First Edition. Patricia de Winter and Peter M. B. Cahusac.
© 2014 John Wiley & Sons, Ltd. Published 2014 by John Wiley & Sons, Ltd.
Companion Website: www.wiley.com/go/deWinter/startingstatistics

unlucky collecting the particular set or data, and they were not truly representative – even though we sampled randomly). In order to quantify that uncertainty, we need to know something about probability. Probabilities vary on a
scale from 0 (impossible, e.g. by this evening I will be sunbathing in a deck
chair on Mars) to 1 (definite, e.g. the sun rising tomorrow morning). The first
of these probabilities is not actually 0, nor is the second 1, but they approach
these values. In the middle is .5, or 50%, which is the probability of obtaining
a head, say, if we were to toss an unbiased coin. To convert from a probability
to a % you need to multiply by 100. Getting a 6 throwing a die has a probability of 1 in 6 (1 divided by $6 = 1 / 6 = .17 = 17\%$, approximately). We divide
the number of outcomes we anticipate by all possible outcomes. If we were
interested in throwing either a 6 or a 1, then that is two out of six possible outcomes. That means the chances of getting that result would be $2 / 6 = .33$. If,
however, we want to calculate the probability of getting that result on the first
throw **and** the second throw, we do **not** just add .33 to .33, we have to multiply
them together: $.33 * .33 = .11$. The probability of two separate events both
occurring will always be smaller than any one event alone. So the probability
of throwing a 6 or a 1 on four separate throws is even more unlikely, it is .33
multiplied by itself four times, or $.33 * .33 * .33 * .33 = .012$. That's a rather
cumbersome way of writing it out – we can express it more simply as $.33^4$.
Probability is used very widely. For example, we might be told by the weath-
erman that the probability for rain in your region tomorrow would be 30%.
Does that mean there is a 30% chance that it rains in the whole region, that
30% of the region gets rain and 70% none, or that it rains 30% of the time?
It specifically means that any specified point in the region has a 30% chance
of getting rain. The probability of my horse winning in a 40 horse handicap
race like the Grand National, might be $1 / 40 = .025$, or 2.5%. The bookies
might consider the horse a favourite and give the *odds* against it winning as
7:1. This translates into a probability of winning as $1 / (1 + 7) = 1 / 8 = .125$.
Of course this does not represent the true probability, only the amount that
the bookmaker is prepared to pay out ($7 for a $1 stake) if the horse won, and
after the bookmaker factors in a margin for profit.

3.3   Frequentist probability

In the long run we're all dead

J.M. Keynes

Consider the independent samples experimental design illustrated in Figure 2.2: 10 participants randomly divided into two groups; one group receiving
tea tree oil, the other receiving placebo. The five participants who receive tea
tree oil end up with no spots, while those receiving placebo still have all their

Does tea tree oil do anything?

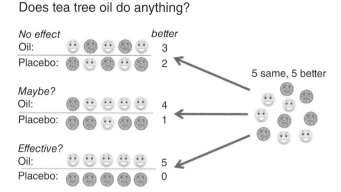

Figure 3.1   Assuming that 5 of the 10 participants get better, then these 5 could be distributed between the two treatment groups in many different ways.

spots. This is just one scenario from such an experiment. There are numerous others. Just one or two could have improved or maybe nearly all, and the distribution of improved versus unimproved could have been different across the two conditions. However, if we imagine that just five got better, then we can work out the probability of all five ending up in the tea tree oil group – **if the tea tree oil did nothing**. For example, if tea tree oil did nothing then we might expect those 5 who got better to be distributed across the two groups: perhaps 2 in one, and 3 in the other, or maybe 4 in one and 1 in the other, and so (Figure 3.1). We need to calculate the probability of all of the five getting better ending up in one group: it could happen, but how likely if neither treatment had an effect?

To do this we need to work out different ways of selecting 5 participants from a group of 10. The particular order in which participants are selected does not matter, so it is the number of *combinations* we need to calculate. We have 10 people. If we make them stand in a row and we rearrange their positions so that each time the combination of where all 10 are placed is different, then there is an exact number of combinations in which we can possibly arrange them. Person number 1 could be standing in position 1, 2, 3, 4, 5, 6, 7, 8, 9 or 10, so could person number 2, 3, etc. Unsurprisingly, there is a mathematical method for calculating the number of combinations of 10 'objects'. It is 10! or, in words, ten factorial; see the Basic Maths for Stats section for more on factorials. So, as we want to know about 5 people who got better out of the 10, that is, the number of combinations of a set of n objects, taken r at a time, we can use the following formula for the binomial coefficient:

$$^nC_r = \frac{n!}{r! * (n - r)!}$$

Most calculators have an ^nC_r button, which you can use to do the calculation.

In our case $n = 10$ and $r = 5$, so the calculation for the total number of combinations is:

$$^{10}C_5 = \frac{10!}{5! * (10 - 5)!}$$

$$= \frac{10!}{5! * 5!} = \frac{3628800}{120 * 120}$$

$$= \frac{3628800}{14400} = 252$$

There are two extreme possibilities here. All five participants who get better are in the tea tree oil group (Figure 2.2, lower scenario), or none of them are (i.e. they would have got better without the tea tree oil, with the placebo). Therefore the probability of either one or the other of these extreme results is 2 / 252 = .0079 (.008 to 3 decimal places). How small is this probability? It is quite a small probability, less than 1 in a 100. It is almost the same probability as getting seven heads in seven consecutive coin tosses ($.5^7 = 0.0078125 \approx$ .008). Try it. Now, if you were to repeat that 100 times, that is, throwing 7 coins 100 times, you would still have only a 54% chance of getting 7 heads.[1] You can appreciate how small a probability .008 is. Now, what does this probability actually mean in the context of this study? We are interested to know whether the data provide evidence that tea tree oil reduces facial spots. All five participants receiving the oil improved, while none of those receiving placebo did. Perhaps those five were going to improve, regardless of what treatment they received. Those five who improved could have by chance been randomly selected to be in the oil treatment group – the probability of that happening was .008. It could happen by chance – but it is unlikely. That probability, which we will call a *p value*, represents the probability of obtaining a result as extreme as ours assuming tea tree oil did nothing. According to accepted statistical practice, we decide to reject the possibility that there was no effect. In effect we are deciding that a fluke did not happen (since .008 is such a small probability). So we conclude that there is evidence that tea tree oil has an effect. Here, we had quite a small *p* value (.008), but let us say that the value worked out to be .10 (10%, 1 in 10), would you then be persuaded of an effect?

---

[1] How did we get that result? The probability of **not** getting 7 heads in a row is $1 - 0.0078125 = 0.9921875$. So not getting 7 heads in a row by tossing 7 coins 100 times, would be $0.9921875^{100} = 0.456430997$. Therefore the probability of getting at least 1 run of 7 heads in a row would be $1 - 0.456430997 = 54.35690026$ or 54.36%.

What we have just done is known as *significance testing* or *null hypothesis significance testing*. We say that something is *statistically significant* if we decide that there is an effect. The statistician Ronald Fisher who devised the significance test expressed the situation as: 'Either an exceptionally rare chance has occurred, or the theory of random distribution is not true' (Fisher, 1990). The meaning of significance used here is specific to statistics and should not be confused with *importance*. The widely accepted criterion adopted for deciding whether there is a statistically significant effect is .05 (5% or 1 in 20). In our study above, we obtained a value much lower than this, which is why we concluded that there was an effect. The procedure is also called null hypothesis testing because we are testing a null hypothesis that there is no effect (null means nothing). Shorthand for the null hypothesis is H_0, and for the alternative (or experimental) hypothesis it is H_1.

The null hypothesis above was that the effect of tea tree oil was no different to placebo in affecting the number of facial spots. Our null hypothesis is actually referring to the **population**, rather than our particular sample of participants. We use our sample to make an inference about the population. If it works in our experiment's sample we assume it would also work in the population. Of course we may be wrong. If there is really no effect in the population, but by chance we happen to observe it in our particular sample, then we have obtained a false positive result and made an error. This is known as a *Type I error*. It would be unfortunate to keep claiming positive results for different treatments, so we try to minimise this type of error to the 5% referred to earlier. This 5% is known as our *significance level,* and the probability is denoted by the Greek letter α. We compare the p values obtained in a statistical test with this probability. If it is smaller than .05 (5%), then we reject the null hypothesis, if it is larger, we do not reject it. Logically there is another type of error where our data do not allow us to reject the null hypothesis. Say, for example, we obtained a p value of .06, but there is actually an effect in the population. We would not be able to reject H_0 (because .06 > .05), but there is an effect present. This is called a *Type II error*, whose probability is denoted by β. The different decisions are tabulated in Table 3.1 which highlights these

Table 3.1   Depending on whether or not the null hypothesis H_0 is true, different types of error can be made (highlighted). If H_0 is true and it is rejected, then we are making a Type I error. Conversely, if H_0 is false and it is not rejected, then we are making a Type II error. Otherwise correct decisions are made

	H_0 true	H_0 false
H_0 rejected	Type I error α	Correct decision $1 - \beta$
H_0 not rejected	Correct decision $1 - \alpha$	Type II error β

Table 3.2 Court decisions are similar to decisions in science. Either the person is innocent or they are guilty (similarly either the null hypothesis is true or not true). The two types of errors are: a false charge of an innocent person and the release of a guilty person. What we really want to do is convict the guilty party, just as in a study we wish to find a statistically significant result (assuming H_0 is false)

		True state of the world	
		Innocent	Guilty
Decision	Convict	α ✗ False charge	$1 - \beta$ ✓ Correct conviction
	Release	$1 - \alpha$ ✓ Vindication of innocence	β ✗ Not proven

errors. So we can think of the Type I error as a false positive, and the Type II error as a false negative.

This decision process differs from significance testing by including two hypotheses and two types of error. This approach, known as *Neyman–Pearson hypothesis testing*, frames the four available choice options in cost–benefit terms, with specific probabilities attached to each outcome. The decision process in an English court of law is similar to this in that a decision must be made to convict or release a defendant. The tabulated decision options are shown in Table 3.2. Here, we specify that the null hypothesis is that the person is innocent of the charge, as we would wish to minimise the probability of a Type I error (the undesirable cost of convicting an innocent person). The Type II error here consists of releasing a guilty person. In Scottish Law this is known as the 'Not Proven' (or bastard) verdict. In significance testing we either reject or do not reject H_0, the latter corresponding to the 'Not Proven' verdict. We also never accept as true either of the hypotheses H_0 or H_1. The Neyman–Pearson approach is consistent in practice (if not in theory) with significance testing in using 5% as our significance level, although Ronald Fisher strongly disapproved of it. The distinction between the two approaches is highlighted by their two concepts: Type I error (α) and the p value. Typically these are assumed to be one and the same thing since they both represent the tail region(s) in a probability distribution and are often associated with a probability of .05. However they are different. The Type I error is specified before a study is done, typically $\alpha = .05$, and refers to the long-run (many studies) probability of making the error of rejecting H_0 when it is true. In contrast, the p value represents the probability calculated from a particular set of data, and is regarded as the strength of the evidence against H_0. Anyway, the distinction should not unnecessarily distract us right now at a practical level, although it may be important at a theoretical level. The Neyman–Pearson approach is particularly useful in that it allows us to specify the probability of obtaining a statistically significant result if the null hypothesis really is false (this is known as the *power* of the test, as will be covered in Chapter 5).

3.4 Bayesian probability

Probability may be described, agreeably to general usage, as importing partial incomplete belief.

Francis Ysidro Edgeworth

Get a cup of coffee. You'll need a little extra attention to follow this section.

Let us say it is known that 1% of people are good. This figure is known as the *prevalence*, in this case of goodness among the population. Is 1% over-optimistic? For the sake of argument let us assume that people are either good or bad, as if it were determined by a gene. Now we have a morality test (Google tells us there is such a thing, but not useful for our purposes), though it is not a perfectly accurate test. The accuracy of the test can be defined in two ways: how often it identifies people who are good (a true positive result), and how often it identifies people as bad (a true negative result). The first of these is known as the *sensitivity*. This is the same as *statistical power*, briefly mentioned earlier, and the major topic of Chapter 5. Let us assume that the sensitivity is 99%, that is, 99 of 100 good people tested will show up with a positive result. That sounds great – we wouldn't want to upset good people with a result that failed to show that they were good (they really are good, you see). What about the bad people and what happens when we test them? This aspect of accuracy is known as *specificity*, and it is the probability of a bad person correctly testing negative (i.e. not good = bad). We also want this to be high because we want those people who are bad to have a high probability of getting a negative morality test result. They are bad, after all, so a negative morality test result is what they deserve. Fine. Let us assume this probability is 96%, so that 96 of 100 bad people will show a negative test result. The sensitivity and specificity are measures of getting the right result for the type of person tested (correct for the good people and correct for the bad people).

Obviously, and regrettably, errors can be made. These can be determined from the 99% and 96% probabilities just mentioned. So, there is a 1% chance of a person who is good testing negative (100% − 99% = 1%). Shame! This means 1 in 100 good people receive a test result informing them that they are bad. From the other probability we get the chance that someone who is actually bad testing positive on our morality test. This will be 4% (100% − 96% = 4%). Horror! Here we have 4 in 100 bad people who show up with a positive morality test result – apparently good, when we know they are actually bad. As you can appreciate, both types of error should be minimised. Which do you think is the worst type of error, and which should we try to minimise most?

These errors are identified with the two types of errors described earlier: 4% corresponds to the Type I error (*false positive*) and 1% corresponds to the Type II error (*false negative*). By assigning the types of error in this way we are assuming that the worst error is for a bad person to test positive (i.e. to appear to be a good person).

Let us select a person **at random** from our population. Give them the morality test and suppose for this particular person the result of the test is **positive**. That is, it suggests that the person might be good. But what is the actual probability that they really are good, **given** a positive test result? Remembering that the prevalence of good people is 1% (see the start of this section), we can accurately calculate this probability. It's not difficult, as you'll see. We must imagine an idealised sample of 100 people. Because it is idealised it will have 1 good person in it, remember only 1% are really good. It is highly likely that that person will test positive, since the sensitivity is 99%. For the sake of our argument we will assume that the good person did indeed test positive. Of the 99 remaining people who are all bad, 4% of them, near enough 4 people (actually $.04 * 99 = 3.96$), will test positive even though they really are bad. These are false positives. The total number of people that tested positive in our 100 idealised sample will be $1 + 4 = 5$. However, only one of these is actually good. This means that the probability of being good after testing positive is $1 / 5 = 0.2 = 20\%$. So, despite quite an accurate test (99% sensitivity and 96% specificity), the probability that someone is good given a positive test result is actually quite modest. The 20% probability is known as the *posterior probability*, because this is the probability calculated **after** the test result is obtained. There is an 80% (100% – 20%) chance that the person is actually bad.

This unexpected result comes about because the prevalence for good people is so low, only 1%. The prevalence is also known as the *prior probability*. This calculation has been done using the Bayesian approach. Another way to view this calculation is to use a decision tree in Figure 3.2. At the ends of the decision tree, to the right, are shown the probabilities of obtaining each of the four possible results: true positive, false negative, false positive and true negative. Because no other results are possible, these four probabilities add up to 1. Two of the probabilities are where the test was positive (true positive = .0099, false positive = .0396), adding those together gives us our probability of obtaining a positive test result = .0495. The true positive probability is .0099, so if we divide this value by the probability of obtaining a positive result we get $.0099 / .0495 = .2$.

Another way to work this out is to use *odds*. The odds of testing positive depending on whether the person is good or bad is $(99 / 4){:}1 = 24.75{:}1$. This tells us that the test is 24.75 times more likely to test positive if the person is good compared to if they are bad. This is known as the *likelihood ratio*. It is also known as the *Bayes factor*. The *prior odds* are 1:99, so multiplying this by our likelihood ratio, we obtain the *posterior odds* for the person being

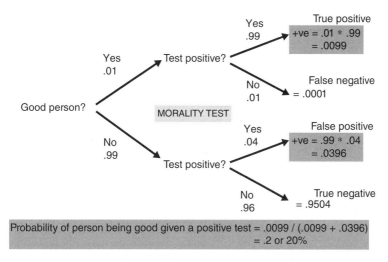

Probability of person being good given a positive test = .0099 / (.0099 + .0396)
                                                       = .2 or 20%

Figure 3.2 Decision tree to visualise probabilities in our testing scenario. At the left side, people divide into either good or bad, with probabilities of .01 and .99, respectively. At the next stage, people divide depending on whether they test positive for the goodness test. If the person is good (top half), then there is a 99% chance that the test will prove positive. If the person is bad (bottom half), then there is a 4% chance that they will test positive. The four different outcomes at the right are labelled True/False and positive/negative, and probabilities are given for each outcome. The total of these probabilities adds up to 1 (100%).

good given a positive test (24.75 / 99):1 = 0.25:1 or 1:4. We convert this to a probability by calculating 1 / (4 + 1) = .20 = 20%.

What happens if we approach this issue from a hypothesis testing approach? This is clearly artificial because we would not normally take a single result from one person to do a hypothesis test. However, we have probabilities which allow us at least to view it from a frequentist angle. Our positive test result tells us that the probability of obtaining this result is less than 4% assuming that the person is bad. The 4% is our probability of getting such an unlikely result given the null hypothesis is true (in this case that the person is bad). As this is a small probability, and less than our significance level of 5%, we should reject the null hypothesis and conclude that the person is good.

So here we have different approaches telling us to make opposite decisions: Bayesian tells us the person is bad (80% chance), while the significance test tells us the person is good (reject null hypothesis of person being bad). Which is right? Well, here the Bayesian approach is using more information (the prevalence or prior probability), and it is right in this situation. The frequentist approach only uses the specificity of the test to guide its calculation and therefore ends up with an inadequate and misleading answer.

If the prior probability is .5 (half the population are good), then the two different approaches agree more closely (and they agree precisely if the sensitivity and specificity values are identical). To show this we can use the odds form of calculation. The likelihood ratio would be (96 / 4):1 = 24:1. The prior odds for .5 is 1:1, so multiplying with the likelihood ratio we get 24:1, which is the same as 1:0.04167. The probability then works out to be 1 / (0.04167 + 1) = .96.

However, the conclusions of their analyses are phrased differently. The Bayesian says there is a 4% probability that the person is bad given the positive test, while the frequentist would still say there is a 4% probability of obtaining this positive test assuming the person is bad. With both approaches it would be reasonable to conclude that the person is actually good (they tested positive). With Bayes you can quite meaningfully talk of a 96% probability that the person is good, given a positive test. More awkwardly, the frequentist can state that there is a 96% probability of testing negative, assuming the person is bad. This frequentist statement doesn't further enlighten us, and is actually very confusing, especially since the person actually tested positive. Had the person tested negative, then the Bayesian would be able to say that there is a 96% probability that the person is bad (given the negative result). The frequentist would simply not reject the null hypothesis that the person was bad because the negative result was consistent with the null hypothesis, and would simply repeat the above saying that there was a 96% probability of getting that result assuming the person is bad.

From a frequentist approach, you should **never** be tempted to say: 'There is a 96% probability of the person being good', when it should be: 'There is a 4% probability of obtaining this positive test assuming the person is bad'). The first statement is known as the fallacy of the *transposed conditional*. Be clear that the probability is concerned with the test result, not the person being good or not, but is conditional on the person being bad (the null hypothesis). You can see that the Bayesian approach actually gives us what we want: the probability of the hypothesis being true or false (person is bad or good), given the data (the positive or negative test result). Both approaches give us conditional probabilities, that is, probabilities that are conditional either on a test result (Bayesian) or on a hypothesis (frequentist).

The Bayesian approach is useful here, and it is increasingly used in diagnostics and expert systems (e.g. detection of spam emails) and medicine (McGrayne, 2012). However, the Bayesian approach becomes more problematic and complicated when applied to hypothesis testing of data obtained in many scientific research areas. This is because we often have no idea what the probability of H_0 or H_1 being true is **before** the data are collected. That is, we rarely know what the prior probability is. The frequentist approach does not need to specify this. However, if previous data have been collected, then the results of those data can be used to indicate what the prior probability might be, in ball park terms. Moreover, the prior probability can be repeatedly updated in this way, refining and pinpointing more precisely the posterior and hence the new prior probability with each cycle: prior probability > collect data > calculate posterior probability > update prior probability >

collect more data … etc. As the evidence accumulates, its weight determines the posterior probability, even if the prior probability is discrepant with it. However, it still remains contentious that in situations where no prior information is available, different people assign different prior probabilities, and accusations of subjectivity begin to fly. Significance testing, as it stands, is reasonable enough, although there are subjectivity problems there too (Simmons *et al.*, 2011). This is especially the case when we are performing multiple statistical tests on a set of data, or deciding when to stop collecting data. It is possible to obtain, by chance, statistically significant results in situations where a priori the null hypothesis is very likely to be true (e.g. in studies of precognition, see Wagenmakers *et al.*, 2011). The Bayesian approach is often more conservative and guards against reaching misleading conclusions on the basis of 'extraordinary claims require extraordinary evidence'.

3.5   The likelihood approach

The likelihood ratio was mentioned above as the objective part of the Bayesian approach. This represents the ratio of likelihoods: that is, the probability of obtaining the evidence given the alternative hypothesis relative to the probability of obtaining the evidence given the null hypothesis. The *likelihood approach* (Edwards, 1992) employs likelihood ratios and is an alternative approach to those already mentioned.

A very simple example will give you some idea about the logic of this approach. Imagine we have two large tubs of smarties (coloured chocolate discs), one tub is black the other white. The black tub contains an equal number of blue and yellow smarties, while the white tub contains one quarter blue smarties and three quarters yellow smarties. We flip a coin to decide which tub to pick smarties from, so each tub has an equal chance of being selected. We are interested in deciding which particular tub had been selected by the coin toss. We do this by considering the subsequent evidence gathered (smarties picked). So we now pick out three smarties (without looking into or at the tub, obviously!). Let us say all the smarties picked are blue. We can then calculate the *likelihood* of obtaining this result for each tub. The likelihoods in each case are obtained by simply multiplying the probabilities of each of the three events together. For the black tub the likelihood of obtaining 3 blue smarties is $1/2*1/2*1/2 = 1/8$. For the white tub the equivalent calculation is $1/4*1/4*1/4 = 1/64$. The likelihood ratio is then the ratio of these probabilities. The ratio in favour of the white tub is $(1/64)/(1/8) = 1/8$, while the ratio in favour of the black tub is the reciprocal of that, 8. In the likelihood approach, a likelihood ratio of 8 (or $1/8$) is regarded as 'fairly strong' evidence. Let us say we picked another two smarties from the tub and both were again blue. We can recalculate our likelihood ratio to give us $(1/4)^5/(1/2)^5 = 1/32$ in favour of the white tub and 32 in favour of the black tub. This is regarded as 'strong' evidence in likelihood terms. Now, let us say that a further smartie is selected, but this time it is yellow. We can recalculate our likelihood, but remember as mentioned earlier the likelihood is calculated by simply multiplying the probability of each of our obtained events together, so we do not need to consider the ways of obtaining the result by the number of combinations (as we would if we were calculating the **probability** of obtaining five blue smarties and one yellow smartie from six selected for each tub). This gives us the following $((3 / 4) * (1 / 4)^5) / (1 / 2)^6 = 0.046875$ in favour of the white tub and 21.333333 in favour of the black tub. Notice that this new event, one yellow smartie has now reduced the likelihood ratio from 32 for the black tub, although the evidence still remains stronger than 'fairly strong'. If we picked out twice the number of smarties with the same ratio of blue to yellow smarties (i.e. 10:2), then our likelihood ratio in favour of the black tub would be just over 455 (and 0.002 for the white tub).

We can continue collecting data like this, building up evidence without worrying about multiple testing as we do with the significance testing approach. As we collect more data the likelihood ratio will continue to increase and better inform us about which tub is the most likely one that we are selecting our smarties from. Now, substitute hypotheses for tubs, and we are talking about the ability to compare hypotheses according to the data collected. It is all about which hypothesis is supported best by the actual data.

There are many features which should make the likelihood approach the preferred method of inference for judging the strength of evidence from data (Dienes, 2008). First, it compares directly the evidence from competing hypotheses, so that their relative evidence is easily interpreted as evidence for or against a particular hypothesis. Second, the evidence generally gets stronger, for or against a particular hypothesis, the more data that is collected (whereas in hypothesis testing the probability of a Type I error remains constant, regardless of how much data is collected). Third, it uses only the evidence obtained, and not what could have been obtained (unlike statistical testing which gives the probability of a result as extreme or more extreme than the observation). Fourth, it is relatively straightforward to combine likelihood ratios from different studies (as might be done in a meta-analysis) to increase the strength of the evidence. Fifth, it avoids the subjectivity of the Bayesian approach and the inconsistencies and subjectivity of the hypothesis/significance testing approach (Dienes, 2008). It seems likely that this approach will gain wider acceptance in years to come.

3.6   Summary

- Common notions of probability were introduced and how they are represented as a value from 0 to 1, a percentage or as odds.

- The **frequentist approach** is the dominant form of statistical testing used in many areas of science. It assumes long-run probabilities.

- Within the frequentist approach there are two dominant procedures. **Significance testing** requires reporting the p value obtained by a statistical test. The smaller this value is, the greater the evidence against the null hypothesis H_0. The probability represents the chances of obtaining the data (and more extreme) assuming H_0 is true. The other procedure, **Neyman–Pearson hypothesis testing** considers the alternative hypothesis H_1, as well as the two types of decision error that can be made following a statistical analysis.

- The Bayes approach more directly attempts to determine the probability of a hypothesis. This is done if the prior probability (or prevalence) is known. It is possible to calculate the probability of a hypothesis given the data collected (i.e. the inverse to the frequentist approach). It is particularly useful in diagnostic settings where the prevalence of an event (e.g. disorder) is known.

- Both approaches have difficulties and have subjective elements.

- Finally, the likelihood approach is briefly mentioned. This has important advantages over the other approaches and may become more popular in future.

References

Dienes, Z. (2008) *Understanding Psychology as a Science: An Introduction to Scientific and Statistical Inference*. Palgrave Macmillan.

Edwards, A.W.F. (1992) *Likelihood*. Johns Hopkins University Press.

Fisher, R.A. (1990) Statistical methods and scientific inference. In: Bennett, J.H., editor. *Statistical Methods, Experimental Design and Scientific Inference*. Oxford: Oxford University Press, p. 42.

McGrayne, S.B. (2012) *The Theory That Would Not Die*. Yale University Press.

Simmons, J.P., Nelson, L.D. and Simonsohn, U. (2011) False-positive psychology: undisclosed flexibility in data collection and analysis allows presenting anything as significant. *Psychological Science*, 22(11): 1359–1366.

Wagenmakers, E.-J., Wetzels, R., Borsboom, D. and van der Maas, H.L.J. (2011). Why psychologists must change the way they analyze their data: the case of psi: comment on Bem (2011). *Journal of Personality and Social Psychology*, 100(3):426–432.

4
Summarising Data – 'Transforming Data into Information'

4.1 Aims

This chapter is divided into two parts, describing data numerically and summarising data graphically. We will first explain that research, whether observational or experimental, often generates a large amount of data that needs to be summarised in some way so that we can gain interpretable information from it. The first half of the chapter will focus on graphical representations of symmetrical and asymmetrical data distributions using histograms and describing data numerically (descriptive statistics). The second half will discuss ways of summarising data graphically using descriptive statistics (summarised) rather than raw data. We will also deal with different types of plot and what they tell us.

4.2 Why summarise?

'The goal is to transform data into information, and information into insight'. This statement is attributed to Carly Fiorina, an American Business Executive and was made in relation to business but describes exactly the broad aim of Statistics. First, we design an experiment or observational study and amass data, sometimes very large amounts of it and increasingly so in this age of genomics and increased computing power. These raw data are numbers, words, images, from which we intend to extract information and detect

Starting Out in Statistics: An Introduction for Students of Human Health, Disease, and Psychology
First Edition. Patricia de Winter and Peter M. B. Cahusac.
© 2014 John Wiley & Sons, Ltd. Published 2014 by John Wiley & Sons, Ltd.
Companion Website: www.wiley.com/go/deWinter/startingstatistics

Table 4.1   Data from an experiment to determine the effect of three treatments on renal excretion rate as mL/h. For practical reasons, groups of five subjects have been tested on different occasions but is not clear which observations belong to which groups

85	68	84	87	72
100	91	122	99	88
55	56	51	49	52
64	63	62	74	66
97	98	88	93	84
61	52	39	66	61
83	72	99	88	84
89	91	94	106	110
79	64	50	50	58
93	79	68	68	86
92	95	95	86	88
40	49	43	59	38
89	79	92	73	83
103	104	96	82	127
59	63	55	50	40

a pattern, if it is present. We use a statistical test to determine whether a pattern could exist within our data; whether a factor has had an effect or one group reacts differently from another, but all we have initially are rows and columns of unprocessed data. We need to transform our data into information, or summarise it in some way so that we can visualise patterns within it. Humans generally are very good at processing visual data – very few people can look at a sheet of numbers and summarise patterns within it mentally. As an example, take a look at the data in Table 4.1, which is a fairly small data set. A pattern or trend is not immediately obvious. In Table 4.2, the data have been labelled and rearranged so that values from one group are together in a block. In this arrangement it is possible to observe that the treatment 2 values appear to be a bit higher than the others; for a start there are some numbers that exceed 100. Although this helps somewhat, we still have 75 numbers to digest so it's still not that easy to see what is going on, if anything, in these

Table 4.2   Data from Table 4.1 rearranged and with row and column headers added. Each observation is from one individual – there are 25 subjects per treatment group (75 in total). Fifteen different subjects, divided into groups of five per treatment were tested on each occasion

Test occasion	Treatment 1					Treatment 2					Treatment 3				
1	85	64	83	93	89	100	97	89	92	103	55	61	79	40	59
2	68	63	72	79	79	91	98	91	95	104	56	52	64	49	63
3	84	62	99	68	92	122	88	94	95	96	51	39	50	43	55
4	87	74	88	68	73	99	93	106	86	82	49	66	50	59	50
5	72	66	84	86	83	88	84	110	88	127	52	61	58	38	40

data. What we really need is to arrange the data into a graph, to summarise them so that we don't have to try to interpret a sheet of numbers. We'll start with a very simple type of graph called a dot plot (Figure 4.1). Each dot on the graph represents one of the numbers in Table 4.1. Now we can tell at a glance that the values of renal excretion rate for treatment group 3 are all lower than those for treatment group 2 – there is no overlap between the two groups, but the values for treatment group 1 overlap with those of both other groups. It is easier and quicker to determine a pattern from the graph than from the table.

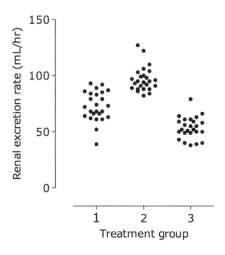

Figure 4.1   Data from Table 4.2 converted to a dot plot.

A dot plot may work well with small numbers of observations per group, but if we have many observations the graph will become rather messy. What we therefore need to do is to summarise our data numerically first and then plot it. In order to do this we will use some *descriptive statistics*. These are numbers that describe certain characteristics of a set of data, so that they can be summarised, and plotted on a graph to visualise them. After all we are not usually interested in the results for each individual whether they be people, tissues, cells, or genes, what we really want to know is if groups of people or cells etc., all respond in a similar way when subjected to an experimental treatment.

4.3   Summarising data numerically – descriptive statistics

4.3.1   Measures of central location

The most common and a very effective way of summarising many numbers using a single value is to use a measure of central location. This gives us an estimate of the value that represents the middle of the data. The 'middle' here means the central value when the observations are organised from lowest to highest (ranked) around which most of the other values occur. The most widely used measures of central location are the *mode*, the *median* and the *arithmetic mean*, often shortened simply to the *mean* and given the symbol $\bar{x}$ (pronounced x-bar). These are all types of *average*. The word average is not useful as it can mean any of these measures of central location, and indeed some others that are beyond the scope of this book. Unfortunately, Microsoft Excel uses the word 'average' to denote the arithmetic mean, which can be confusing. It is important, therefore, to always specify the type of average when reporting measures of central location. So the first things we wish to know is where the centre of the data lies and which measure of central location we should use.

The mode is simply the value or class interval that occurs most frequently in a data set. The mean is calculated by adding up all the observations and then dividing by the number of observations (the number of observations is symbolised by n for short). The median is the middle value, or that which divides the data in exactly half after the observations have been ranked from lowest to highest. Where the number of observations is even, the median is calculated as the mean of the two middle values.

Why do we need different measures of central location? Well, to answer this question it is necessary to talk about data distributions. In order to do so we will first use a non-biological generic example that should be readily understood by anyone, irrespective of their field of study. This example is annual salaries. We will use the United Kingdom as the geographical location. Most people in the United Kingdom earn a modest salary; they are neither very rich nor very poor. There are many people earning low to moderate salaries and a fair number earning a comfortable salary. But the United Kingdom is also home to a sizeable number of people who are very wealthy indeed and earn at least 10 or even 20 times more than the people that they employ. And then there are also the super-rich. If we were to calculate the mean of all UK salaries (full-time and part-time employees), we would find that it is higher than the median. Statisticians call this a skewed distribution. We can roughly determine the shape of a distribution by plotting the data on a graph called a histogram and we can use this information to assist us in deciding which measure of central location we should use to find the middle of the data. First, we sort the data into groups representing a range of salaries, for example, <£5000; £5000–14,999; £15,000–24,999; £25,000–34,999; £35,000–44,999 etc. These groups are called class intervals. Next, we count how many people earn the range of salaries in a class, that is, the *frequency* of occurrence (Figure 4.2). Now we plot the histogram with class intervals on the *x*-axis and frequency on the *y*-axis. Note that instead of using the range of each class interval on the *x*-axis as we have done, we could simply use the midpoint of each range, which is more common, that is, 2500; 7500; 12,500 etc. A skewed

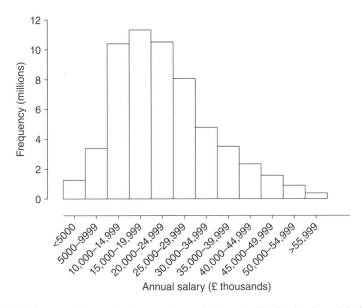

Figure 4.2   Histogram of approximate annual UK salaries. The distribution is positively skewed.

distribution is not symmetrical and has a 'tail' either to the left or to the right. In the case of UK salaries the tail is to the right, because it is the very high salaries that are skewing the data. The arithmetic mean is greater than the median. If we were to plot a histogram of mini mental scores in people aged over 60, we would likely find a negatively skewed distribution (tail to the left) as quite a few people will have low scores.

So our next question is how do we know where the middle of the data lies by plotting a histogram? How does this skewness in data help us to understand why we use different measures of central location? Well, if we calculated the mean full-time UK salary from the raw data it would be £31,500, whereas the median would be lower at £26,300. This is because the mean is influenced by the salaries of the higher earners pushing its value upwards, whereas the highest salaries have no influence on the median, it being the exact middle point of the data. So where data are skewed in either direction, the mean and the median differ, and it is the median that provides a better measure of central location as the effect of extreme values is small. The mean in these cases will provide a falsely high or low estimate of the 'average' value.

The mode is perhaps the least informative measure of central location as it is simply the value or class interval that occurs most frequently. Sometimes it may coincide with the mean and/or the median, and at other times it may be completely different from them. For example, if we were to take the fasting blood glucose of 145 subjects who attend a clinic for the investigation of urinary frequency (passing urine often), we might find that the mode is an abnormally high value of 7.5 mmol/L (Figure 4.3). Most blood glucose values

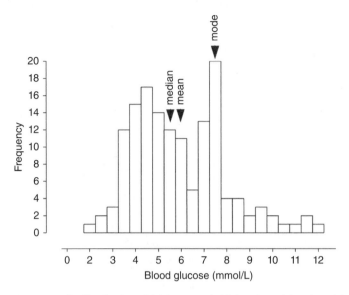

Figure 4.3 Histogram of a distribution which has a mode higher than either the arithmetic mean or median. There are 20 subjects with a fasting blood glucose between 7.25 and 7.74 mmol/L represented by the class interval midpoint modal value of 7.5 mmol/L. The median class interval midpoint is at 5.5 mmol/L and there are only 12 subjects with a blood glucose value between 5.25 and 5.74 mmol/L.

occur around 4–6 mmol/L and are within the normal range of fasting levels and the median is located at 5.5 mmol/L (the distribution is skewed with a mean value 5.9 mmol/L, so the median is a better measure of central location here). The mode differs greatly from the median in this case, because it reflects the presence of a disease group of previously undiagnosed diabetics. While there are many causes of urinary frequency, including an enlarged prostate in men, an infection in the bladder and neurological bladder problems, a proportion of adult patients presenting with this symptom will have type II diabetes and hence a higher than normal fasting blood glucose. The mode can sometimes provide us with useful information about our data so although not necessarily the most representative measure of central location, it has a place in data analysis. In this example, it may be better to separate the data for the diabetics from non-diabetics.

There are many instances where biological data might exhibit skewness. One example is the expression level of genes, which exhibits a skewed distribution. There are some genes that are expressed at very high levels indeed, the majority are moderately expressed and some have very low expression levels. The data distribution for gene expression is similar to that of London salaries in that they are positively skewed.

Which measure of central location is best for data that are not skewed? Well let's take a look at some more data. Again, we will use an example that can

5'0"   5'2"   5'4"   5'6"   5'8"   5'10"   6'0"

Figure 4.4 Cartoon based on idea of Joiner's (1975) living histogram illustrating a symmetrical distribution of the heights of a sample of female college students.

be readily understood by all: adult human height. As there are gender differences in height (males are generally taller than females), we'll use adult female height first and plot its distribution on a histogram (Figure 4.4). The idea for the cartoon histogram originates from Brian Joiner's (1975) paper in which he used real university students to produce 'living histograms'. There are relatively few very short students on the left side of the x-axis, most students' heights are grouped around the middle of the axis and there are few very tall female students on the right of the x-axis. Note that unlike the previous variables (salaries and fasting blood glucose), the histogram for adult female human height does not exhibit a long tail in either direction. The histogram looks quite symmetrical. If you think about it, the histogram likely reflects your own real life experience; think about how many very short (<5 feet) or very tall (>6 feet) women you observe, say on your way to work, or when shopping in the supermarket. And how many women are closer to 5′4″ in height? If we calculate the mean and median for this variable, female human height using the data from the 1975 histogram of American university students, we would find that they coincide (rounding to the nearest half inch, they are actually both 64.5 inches or 5 feet and 4 and a half inches). When data are symmetrically distributed around a mean that coincides with the median, with frequencies that form a pattern that looks like a bell-shaped curve, they are said to conform to a normal distribution. The normal distribution has played a very important role in the history of Statistics as it forms the basis of tests known as parametric statistical tests, which we will discuss in more detail later.

To return to descriptive statistics, we have demonstrated that selection of an appropriate measure of central location is very important if we wish to accurately describe where the middle (average) of the data is located. The mode may coincide with the mean and median when data are normally distributed but is not the most useful measure. The median best describes the middle of the data when distributions are skewed (not normally distributed), and either the mean or the median may be used when data distributions are normally distributed. A normal distribution is symmetrical and resembles a bell-shaped curve, but not all symmetrical distributions are normal. Figure 4.5 represents a symmetrical distribution that is uniform, rather than normal. A uniform distribution occurs where the frequency is similar across all class intervals, producing a histogram that looks more like a rectangle than any other shape. An example might be a disease that occurs with equal frequency in all age groups. Many diseases exhibit a peak at particular ages, some occur most frequently in children whilst some others are very rare in the young and common in old age. Such diseases would exhibit skewed or normal distributions, but something like the common cold, which does not discriminate by age, might produce a histogram that is roughly uniform, rather like Figure 4.5.

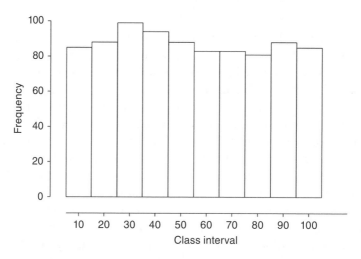

Figure 4.5 A uniform distribution. The frequency of occurrence does not differ much across most of the class intervals so the histogram has more of a rectangular shape rather than curve with a distinct peak.

Another example of a distribution that can be symmetrical but not bell-shaped is one that is *bimodal*. A bimodal distribution has two frequency peaks separated by a dip. An example of when one might occur is where a variable is measured in both males and females and not separated by gender, although gender differences are present. For example, men are, on average, physically stronger than women. An example of this is maximum voluntary contraction (MVC) of muscle. This is typically measured by asking subjects to squeeze an instrument called a hand grip dynamometer as hard as they can, and this measures the force in kg generated by the hand, wrist and forearm muscles. If we plot MVC values for both genders as a histogram, we obtain a bimodal

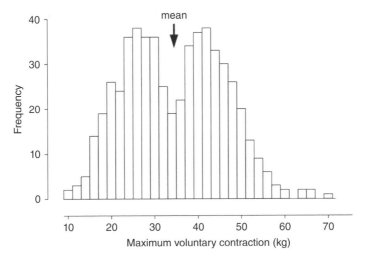

Figure 4.6   A bimodal distribution of maximum voluntary contraction in men and women aged 18–40. The frequency exhibits two peaks – one at around 24–30 kg (the MVC for most women) and the other at around 38–44 kg (the MVC for most men). The mean is 34.6 kg.

distribution (Figure 4.6). The mean for this sample is 34.6 kg, located in the dip between the peaks. This is clearly not a useful description of where most of the observations are located, but the problem is that the observations occur most frequently at the two humps in the histogram and we can't have two measures of central location for one sample. The sensible thing to do here is probably to separate the observations into two groups, one for each gender and we would find that each sample is normally distributed. Now we can use the mean for each gender as a valid description of central location.

4.3.2   Measures of dispersion

The measure of central location is one important descriptive statistic that provides some very useful information about our data, but on its own it is not sufficient to fully characterise them. Another very important property of a set of data is how variable are the observations. It is of critical importance when evaluating our scientific data, because it is the degree of variability of the data that permits us to make inferences about whether we have a real effect or one that is simply due to chance, that is, random variation in a population.

So what do we mean by the variability of a data set? Well, think back to the median UK salary which we observed was £26,300. If we report only this number (as the media often do) it will not give us information about the different levels of salary that people earn – do most people earn salaries between £20–30,000 or between £10–40,000? The former is less variable than the latter. If we are fortunate to have a histogram of the data distribution, we may be able to make a reasonable estimate in order to answer this question but if we don't how else might we describe the data to give us some idea of how much less or more than the median people in the United Kingdom generally earn? Statisticians call numbers that describe the variability of a set of data, *measures of dispersion*. The simplest measure of dispersion is the range over which the data are spread. The range is the difference between the highest value and the lowest value. That's easy to work out but is it an informative measure? The trouble with the range is that when most of the observations are clustered tightly around the centre of the data, but there are a few extreme values, the range will give a falsely large value of variability. This is illustrated in Figure 4.7. The first histogram depicts the age at diagnosis of a disease that affects mostly young adults (Figure 4.7a). The range of the data for this

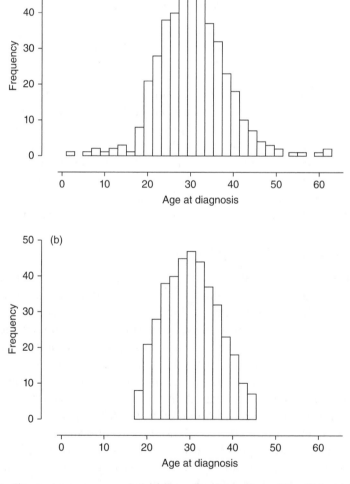

Figure 4.7 The range as a measure of dispersion often does not provide sufficient information. Most of the values in both histograms occur between 18 and 44, but the range of (a) is much greater than of (b).

histogram is 60 (i.e. 62 – 2), but note that most of the values occur between the ages of 18 and 44 with diagnosis rarer in children and older adults. If the few cases at the extremes were disregarded the range would be much smaller at 26 (i.e. 44 – 18, Figure 4.7b). Unfortunately, as the range is, by definition, the difference between the two most extreme values we cannot simply ignore ones that we arbitrarily decide we think shouldn't be included. Although the range provides us with some information, we really need a better way of describing the variability of the data.

A more representative way to describe dispersion would be to use a method that represents the variability of the data in a standardised way. What if, for example, we used the two numbers that include a certain *percentage* of the data around the mean or median value? Whichever numbers comprise the data, large ones, small ones, negative or positive, the percentage would remain fixed. To demonstrate such a method, let us use an example in which data sets have the same central location but differ in dispersion. The data are simply numbers that have been generated by computer software to have an arithmetic mean of 30 and median of 30 and arranged to be symmetrical. They differ however in their dispersion (Figure 4.8). The topmost set is dispersed over a narrower range of values than either the middle or bottom set. The middle and bottom set have an identical range, 69.5, as well as the same central location, but the two histograms clearly look different. This is where the standardised method of dispersion, we will now describe, comes in handy, as it can distinguish between all three data sets.

In order to explain the method we will use a set of nine observations, so that the calculations do not become too onerous: 20, 23, 24, 28, 30, 32, 35, 38, 40. The arithmetic mean and the median are both 30.

Step 1.   Subtract each observation from the mean value:

$$20 - 30 = -10$$
$$23 - 30 = -7$$
$$24 - 30 = -6$$
$$28 - 30 = -2$$
$$30 - 30 = 0$$
$$32 - 30 = 2$$
$$35 - 30 = 5$$
$$38 - 30 = 8$$
$$40 - 30 = 10$$

The numbers we obtain are called the *deviations from the mean* or *deviations* for short. It would be nice if we could just add up all these deviations and use the number we obtain as a measure of dispersion, but if you have done your calculations correctly the

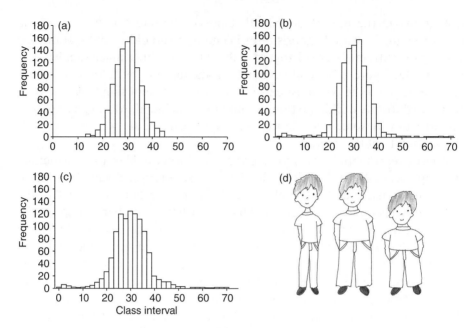

Figure 4.8   Three sets of symmetrically distributed data, each of which has a mean of 30 and median of 30. All three histograms have been plotted on the same scale for comparison. The ranges however differ, for (a) the range is 28 but for (b) and (c) it is 69.5. (a) and (b) look very similar except for the difference in their ranges but (b) and (c) have exactly the same range and central location, and yet histogram (c) is squatter in height and wider around the mean value. If the histograms were people one might describe them as tall and thin, tall and broad, short and broad, rather like the differences between the three figures in (d).

deviations will unfortunately always add up to zero. This is because half the observations will have a negative value when you subtract them from the arithmetic mean. Still, the sum of the deviations is an interesting idea, and in the next step, we'll see if we can find a way round the zero problem.

Step 2.   Square each of the deviations (multiply each number by itself):

$$-10^2 = -10 * -10 = 100$$
$$-7^2 = -7 * -7 = 49$$
$$-6^2 = -6 * -6 = 36$$
$$-2^2 = -2 * -2 = 4$$
$$0^2 = 0 * 0 = 0$$
$$2^2 = 2 * 2 = 4$$
$$5^2 = 5 * 5 = 25$$
$$8^2 = 8 * 8 = 64$$
$$10^2 = 10 * 10 = 100$$

Note that by squaring the deviations, the products are now all positive numbers. Well at least now if we add them up, we won't get zero!

Step 3. Sum the squares:

$$100 + 49 + 36 + 4 + 0 + 4 + 25 + 64 + 100 = 382$$

This number is in fact unsurprisingly called the sum of squares.

Step 4. Divide the sum of squares by the number of observations minus one in the original data set – in our case we had nine so, $9 - 1 = 8$. Obviously a larger sample size will otherwise have a large value simply because it contains more observations, so this step corrects for the sample size

$$382 / 8 = 47.75$$

The number we obtain is called the *variance*. It is effectively the arithmetic mean of the squared deviations (in Step 3 we added up the squared deviations and in this step we have just divided by $n - 1$). That's almost how we calculate the mean of a set of numbers, add them up and divide by n, but here we are using $n - 1$, why? A simple explanation is if we were to use n we would systematically underestimate the population mean (recall from Chapter 1 that we use a sample to make inferences about the population from which it is drawn). This can be demonstrated and we refer you to Grafen and Hails (2002) for the mathematical proof. The number of observations minus one is the number of *independent* observations that contribute to the calculation of the variance. We can think of independent in this way: if we have only one observation we have no information about the variability of a population. If we add another observation, we now have *one* estimate of variability, the difference between the two values. If we add a third observation, we now have *two* estimates of variability; so it is always $n - 1$. The $n - 1$ here is known as the *degrees of freedom*.

So why not stop there? Well, the variance will not be in the same units as our original observations, it's still squared: if we measured something in cm, then the units of the variance will be cm^2, if our measurements were kg, then the variance will be in kg^2. That's a bit hard to get your head around but it's easy enough to fix!

Step 5. Take the square root of the variance:

$$\sqrt{47.75} = 6.910, \text{ or } 6.9 \text{ to one decimal place}$$

We now have a standardised measure of dispersion, 6.9. We have just calculated the *standard deviation* of our data set. We can describe our set of nine numbers as 30.0 ± 6.9, which means the arithmetic mean is 30.0 and the dispersion is 6.9 above and below it. This is short-hand for 30 − 6.9 = 23.1 (one standard deviation below the arith-metic mean) and 30 + 6.9 = 36.9 (one standard deviation above the arithmetic mean). The standard deviation of a sample is given the symbol s.

Let's now go back to our three sets of data in Figure 4.8. There are just over a thousand observations in each so calculating the standard deviations would be a bit tedious if we had to do it step by step. For-tunately, any statistical software or a spreadsheet such as Excel will calculate it for you automatically:

Data set	Arithmetic mean	Standard deviation	Range
A	30.2	5.2	28
B	30.2	7.0	69.5
C	30.2	8.1	69.5

Well, that is interesting; recall that the range for B and C was identical at 69.5, whereas the standard deviations differ, that of C is greater than that of B. This better describes the dispersion we observed in Figure 4.8. We can draw around the outlines of the histograms and superimpose them to get a better comparison of their shape (Figure 4.9). The tall narrow histogram has the smallest standard deviation, 5.2. This is because the data fall within a rel-atively narrow spread around the arithmetic mean value. Some of the values for B are spread out further from the arithmetic mean in both directions, but most values are still fairly close to it and the standard deviation is therefore larger at 7.0. The values for the short wide histogram, C, have the greatest spread and this is reflected in it having the highest standard deviation at 8.1.

The standard deviation is not so-named without good reason. It is a very useful measure of dispersion for symmetrical distributions because

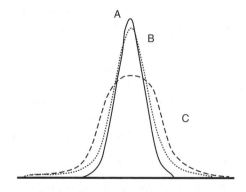

Figure 4.9   Outlines of the three histograms in Figure 4.8: (a) tall and thin, solid line; (b) tall and broad, dotted line; (c) short and broad, dashed line.

mathematically, the values between one standard deviation above and below the mean always comprise approximately two thirds of the observations, and this is what we mean by a standardised measure of dispersion. Where the standard deviation is large relative to the arithmetic mean, we know that the data must be very variable, and where the standard deviation is small compared with the arithmetic mean, we know that the data must be less variable. We now have two numbers that can describe a data set where the distribution is symmetrical.

Will this work with asymmetrical distributions? Well, we know that the arithmetic mean does not best represent the centre of the data for asymmetrical distributions so it follows that the standard deviation may not be the best measure of dispersion. For asymmetrical data, we can use a measure of dispersion related to the median. The median, as we have already observed, divides the data into exactly in half so 50% of values lie below it and 50% lie above it. We can further divide each half so the whole data set is divided into quarters. These are called quartiles (the 25th and 75th percentiles) and this is what we have done for the following set of asymmetrical data (Figure 4.10). The first (or lower) quartile is 18, and the third (or upper) quartile is 92. The second quartile is conventionally given a special name, the median, and here it is 28. The values between the two quartiles comprise the *interquartile range* and this is the measure of dispersion that best describes most asymmetrical data.

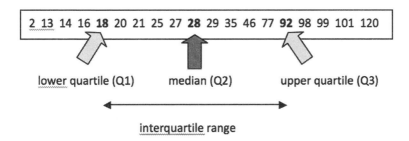

Figure 4.10 The median and lower and upper quartiles of a set of data.

There is a special type of distribution which is asymmetrical, skewed to the right in fact, and this is the *Poisson* distribution. For a Poisson distribution, the data are counts of an event that either occurs or does not (e.g. presence or absence) and that the occurrence of the event is rare. The mean is equal to the standard deviation. An event may not occur, so the lowest occurrence is zero or may occur with decreasing frequency, so the highest occurrence can be any positive number. As an example, suppose that you wished to count the number of eosinophils (a type of white blood cell) in a blood sample. The number of red cells is enormous around 5 billion red cells per mL of blood $(5 \times 10^9/\text{mL})$. This is approximately the middle of the reference range for men and women together (around $3.9–6.5 \times 10^9/\text{mL}$, the reference range varies slightly from lab to lab) as women have lower counts than men. The total white cell count in health is almost 1000 times lower than this (range 4.0–11.0 × 10^6/mL). So one might expect to see one white blood cell in approximately a thousand red cells. Eosinophils constitute only a small proportion of white cells (between 1% and 6%) so we could regard their presence in any particular microscope field a rare event. This means that we will have to examine many fields to find any eosinophils: we would typically apply only a drop of blood to a slide to prepare a smear so there will be only one or two hundred eosinophils on the whole slide. We can imagine, therefore that the presence of eosinophils in microscope fields might exhibit a distribution rather like Figure 4.11. Many fields contain no eosinophils whatsoever (zero occurrence), and where they are present there is mostly only one in a field and less frequently more than one. There cannot be fewer than zero eosinophils in a field so the Poisson distribution is restricted to zero as the lowest occurrence. The mean and the standard deviation for this distribution are both 0.97. In contrast, the median is 1, the lower quartile is zero and the upper quartile is 2.

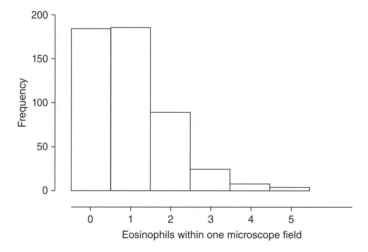

Figure 4.11 A Poisson distribution for the presence of eosinophils in a blood smear. The lowest occurrence of these cells is zero in a field and the highest is five per field. The most frequent events are zero and one per field. Five hundred microscope fields were observed.

4.4 Summarising data graphically

Now that we have established that rather than laboriously trawl through rows and columns of raw data we can simply summarise them using a measure of central location and a measure of dispersion, we can move on to discuss how these numbers could be presented in a graphical format. Experimental studies usually consist of a number of groups given different treatments, perhaps two as in the tea tree oil experiment in Chapter 2 where one group received placebo and the other tea tree oil, or three as in our example at the very beginning of this chapter. However, many groups we have, two, three or more, plotting an appropriate graph permits rapid and very effective communication of our results to the reader. We will first discuss graphs for plotting summarised group data and then we will examine graphs for visualising relationships between variables.

4.5 Graphs for summarising group data
4.5.1 The bar graph

Perhaps the most used and abused type of graph is the bar graph. It is popular because it is simple to construct and easy to understand, but it is often inappropriately used for skewed data and can be very misleading. A bar graph is most appropriate for data that are counts and symmetrically distributed. Let's stick with blood for this purpose and examine the data in Table 4.3. The data are the arithmetic mean and standard deviation for three groups of 12 subjects (6 male and 6 female per group). We can tell from the descriptive statistics that the disease groups have either a lower or higher red blood cell count than the healthy group and that there is a degree of variability in each sample. If we do the maths we can work out one standard deviation above and below the arithmetic mean for each group: healthy 4.48 and 5.32; ulcer 2.63 and 3.77; COPD 5.92 and 7.68 but we can also display these data as a bar chart where the arithmetic mean is represented by the height of the bar and the standard deviation above the mean by the T-shaped line sticking up from the top of the bar (Figure 4.12). At a glance we can visualise the differences between the three samples. Note that unlike a histogram, the bars are separated. This is the correct way to draw a bar graph, whereas in a histogram the bars usually touch against each other. The group names (or categories) are on the x-axis and the count (frequency) is on the y-axis. It's also very important to include information about the sample size and what the error bars (T-shaped lines) represent (in this case the standard deviation). That's a nice clear graph, why can't we use it all the time?

Table 4.3 Red blood cell counts in healthy human subjects and patients with stomach ulcers or chronic obstructive pulmonary disease (COPD), $n = 12$

Condition	RBC count ($\times 10^9$/mL) $\bar{x} \pm s$
Healthy	4.90 ± 0.42
Stomach ulcer	3.20 ± 0.57
COPD	6.80 ± 0.88

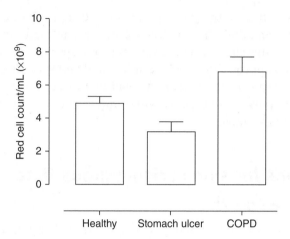

Figure 4.12 Red blood cell counts in health and two diseases. Data are means + standard deviations, $n = 12$.

The problem with the bar graph is that it does not permit visualisation of asymmetry in the distribution of the data and so skewed data can be concealed. This is important because as we will see in later chapters many commonly used statistical tests rely on data that conform to a normal distribution and using them unadjusted for asymmetrical or non-normally distributed data renders the test invalid.

4.5.2 The error plot

The error plot could be regarded as a simplification of the bar graph, where the mean is represented by a dot instead of the height of the bar (Figure 4.13).

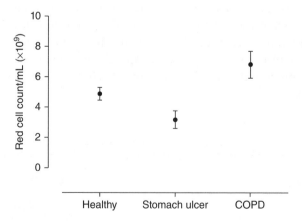

Figure 4.13 Red blood cell counts in health and two diseases graphed as an error plot rather than as a bar graph. Data are means ± standard deviations, $n = 12$.

The error bars here are one standard deviation below the mean and one above it. This plot has provided us with no more information than the bar graph about the red blood cell count data. So are there any other ways in which we can visualise group data without losing important information, such as the distribution? Well, at the very beginning of this chapter we started with a dot plot (Figure 4.1) and concluded that it is problematic for groups containing large numbers of observations as the plot becomes very crowded. An alternative is to use a box plot.

4.5.3 The box-and-whisker plot

Box-and-whisker plots (also known simply as box plots) provide information about the median, the interquartile range, the range excluding outliers (explained below) and the distribution of a sample (Figure 4.14). The range for the data in this Figure is 61 (66 − 5), the interquartile range is 31 (54 − 23), and the median is 35.7. The distribution is not symmetrical as the 'box' is not equally divided into two by the median; the distance between the median and the first quartile is shorter than that between the median and the third quartile. The 'whiskers' are the T-shaped bars that stick up above and below the box and represent the range (lowest and highest values), excluding outliers. Outliers are data points that have extreme values and box plots allow us to visualise these. An outlier is usually any value that lies more than one and a half times the length of the box (interquartile range, IQR) from either the top

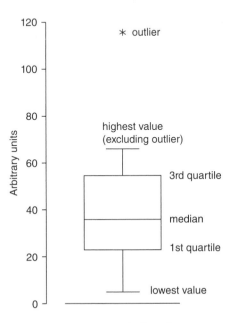

Figure 4.14 Anatomy of a box plot, $n = 30$.

or bottom of the box. Outliers may result from true biological differences, but they can also be due to investigator or experimental error. Their presence will skew the data and increase the standard deviation. In a box plot the outlier values are represented by an asterisk.

4.5.4 Comparison of graphs for group data

In order to see what difference a plot makes, let us now compare data for two groups of 20 subjects with the same mean (25) and standard deviation (5), but different distributions. The data are resting respiratory rate in smokers and non-smokers aged 70 or over. As well as physical fitness a large number of medical conditions can affect resting respiratory rate including, lung disease, heart disease, anaemia, all of which are fairly common in older people, so unless the groups were closely matched to take fitness and disease status into account, which these data are not, one might expect quite a big variation between subjects. Which graph or graphs provide the most useful information about the data here? Well, the least informative is clearly Figure 4.15b, the bar graph. If we were to rely on this graph we might be misled; while the means and standard deviation do not differ between the two groups, their distributions do, as clarified by the dot plot (Figure 4.15a) and box plot (Figure 4.15c). It is rather a matter of opinion as to which of these other two is the best graph to use. The dot plot is fine from the perspective that there aren't too many data points and so they are readily distinguished, but it probably requires some experience to determine that the data for the non-smokers is not normally distributed, whereas that for the smokers likely is. We now also know that the mean is not the appropriate measure of central location for the

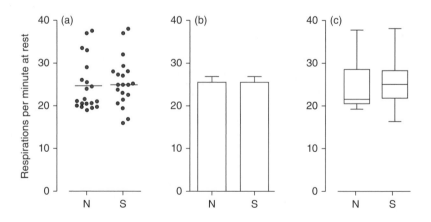

Figure 4.15 Respiratory rate in two groups of people aged over 70, non-smokers, N, and smokers, S, plotted as different types of graph. (a) dot plot, (b) bar graph (c) box-and-whisker plot. Data are means and standard deviations for (a) and (b) and medians and IQR for (c), $n = 20$.

non-smoker group, but conventionally we don't mix means and medians on one graph. It would be better to plot the medians here, which we could do. So we would then have a good idea of central location for both groups but not a clear visual indicator of dispersion, except for the fact that we can see all the individual data points for each group. We could plot the interquartile range as vertical bars, but then it would start to get quite messy. My vote would go to the box-and-whisker plot: at a glance we can tell that the distributions differ and the non-smoker group data are very skewed, we have a measure of dispersion, the interquartile range, and the sample size, 20 is indicated in the descriptive figure legend. Let's suppose that these data appeared in an academic paper you were reading and that the authors had conducted a very common type of statistical test called a *t*-test (which we will meet in another chapter very soon). You would be able to tell just by looking at the dot plot or the box-and-whisker plot that this is an inappropriate test as it is valid only if the data for both groups are normally distributed. The bar graph, however, would mislead you into trusting that the test was appropriate as you have no information about the distribution.

4.5.5 A little discussion on error bars

Graphs for group data are sometimes plotted in academic publications with error bars representing the *standard error of the mean*, or the *95% confidence interval*, which we will discuss in another chapter, and there is much confusion about the difference between the standard deviation and these other two. As the standard error of the mean is always smaller than the standard deviation, some authors like to plot it to make the variability appear lower. The standard error of the mean and confidence intervals provide information about the *variability of the location of the population mean*, and not the variability of the sample. So which error bars you plot very much depends on what you are trying to illustrate: if you want to communicate the variability of the data around the *sample* median or mean, then the IQR or standard deviation should be used. You will come to understand more about error bars as you progress through this book.

4.6 Graphs for displaying relationships between variables

Let's suppose we'd like to know if there is some sort of relationship between two variables. There are two main types of graph for illustrating such relationships: the scatter diagram or plot, and the *x–y* plot. Superficially these two look very similar, but they do differ albeit in a rather subtle way.

4.6.1 The scatter diagram or plot

Scatter plots are used for two situations. One is where there is no dependent variable, hence both variables are *measured*. Let's take resting heart rate and systolic blood pressure in a sample of 40 biology students aged 18–40. We *measure* both variables in each student and plot a graph, but which variable goes on which axis? Well, they are both measured and the resting heart rate is not determined by the systolic blood pressure and vice versa. So it doesn't matter which variable is plotted on which axis (Figure 4.16). We could say that the variables are interdependent (each depends on the other). There will be random error associated with the measurement of both variables. In this case, the random error could be due to factors such as the accuracy and precision of the equipment, health and fitness of the individuals, genetic differences between

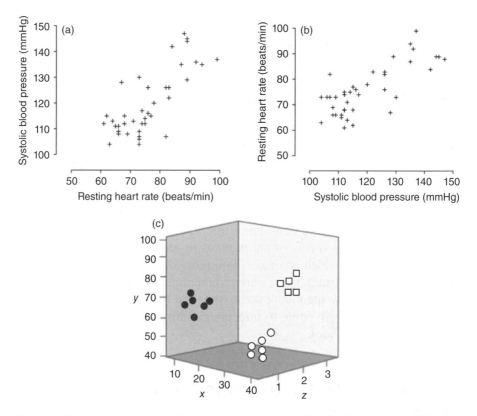

Figure 4.16 Scatter plot of resting heart rate versus systolic blood pressure in 40 biology students aged 18–40. As there is no dependent variable, either variable can be plotted on the y-axis (a) systolic blood pressure on y, resting heart rate on x; (b) resting heart rate on y, systolic blood pressure on x. Data courtesy of Dr. Glenn Baggott, Birkbeck, University of London, used with permission. (c) Three-dimensional scatter plot showing separation of three groups of fictitious data along the z-axis, which would be difficult to distinguish if plotted on two axes.

subjects, experience of the person taking the measurements, etc. The scatter plot may also be plotted with a third variable on a z-axis to produce a three-dimensional graph. This is useful for certain techniques that reduce complex data to a few components, for example, principal components analysis.

In the other situation, only one variable is measured (dependent) and the other is selected by the researcher. The important distinction is the nature of the y-axis variable. Where there is one *measured* variable we call it the *dependent variable* and it must plotted on the y-axis. We say it is the dependent variable because the measurements we obtain for it will *depend on* the values of x that are used. The variable plotted on the x-axis is called the *independent variable* and is not measured but is selected. I will use the example of the energy required to separate the two strands of a stretch of DNA in relation to its sequence. You will likely already know that DNA double helix is made up of four bases: adenine, thymine, cytosine and guanine, abbreviated to A, T, C and G. An adenine on one strand of the double helix always pairs with a thymine of the opposite strand through double hydrogen bonds and cytosine always pairs with guanine but through triple hydrogen bonds (Figure 4.17). It takes more energy to break a triple bond than it does a double bond so we might imagine that if a strand of DNA contains more guanines and cytosines, it will require more energy to force the two strands of the double helix apart (denature the DNA). So we might ask ourselves whether there is a relationship between the variables 'energy required to denature DNA' and 'GC content of DNA'. If we work with DNA a lot, this is something that is useful to know and can be determined by measuring the temperature at which pieces of DNA denature or 'melt' using a technique called real-time PCR (polymerase chain reaction). Molecular biologists call the stretch of DNA a PCR product or an amplicon, because billions of exact copies are made or 'amplified' from the original DNA to obtain enough to measure the melting peak. Note that we are not measuring the GC content; we are *selecting* a number of assays for which we *know* the % GC content (independent variable) and *measuring* the melting temperature (dependent variable). Furthermore, there is no random error in the GC content of the PCR product, the % GC of each is fixed. There is, in fact, a relationship between these two variables (Figure 4.18). The higher the GC content of the PCR product, the higher the temperature required to melt the DNA strand.

Figure 4.17 Base pairing in DNA. A-T form a double hydrogen bond and G-C form a triple hydrogen bond.

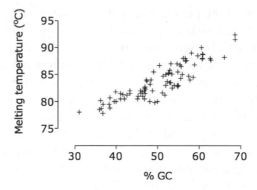

Figure 4.18   Scatter plot illustrating the relationship between PCR product melting temperature and its % GC content, $n = 85$. The melting temperature is the dependent variable; its values are determined by the GC content of the DNA.

4.6.2   The line graph

A line graph is a plot which displays data points connected to each other by a line. It is commonly used to visualise the values of a dependent variable over time, where time is the independent variable. An example might be circadian changes in gene expression. The expression (activity) of certain genes is cyclical over 24 hours and can be measured by determining the number of copies of the messenger RNA molecules in cell or tissue samples. Some genes are highly expressed at night and hardly at all during the day. The expression of other genes is negligible at night and high during the day, some are not affected by the time of day. The gene *clock* (yes, there really is one and it stands for Circadian Locomotor Output Cycles Kaput) is differentially expressed over 24 hours and may be plotted on a line graph (Figure 4.19). There is a decrease

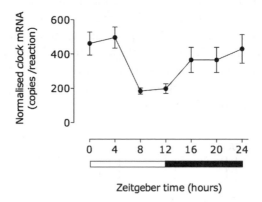

Figure 4.19   Line graph illustrating circadian changes in *clock* gene expression in rat liver. Zeitgeber time 0 is lights on and 12 is lights off. Data are means ± standard deviations, $n = 6$. Data courtesy of Dr. David Sugden, King's College London, used with permission.

in expression during hours of light and an increase during hours of dark. The lines joining the data points imply that the expression of *clock* mRNA is a continuous process.

4.7 Displaying complex (multidimensional) data

Rapid developments in genomics since the publication of the first draft of the human genome coupled with increased computing power have led to the acquisition of enormous data sets, which were previously unattainable. Now, instead of measuring the activity of a few genes at a time, we can measure the activity, methylation status, polymorphisms or mutations of genes across the entire genome (>20,000 genes). Communicating the results of such complex data is challenging, but there are illustrative methods that can achieve this. When analysing the changes in sequence (mutations or polymorphisms) that occur, in a disease, there will of course be many stretches of DNA that are normal and only a proportion will be altered. The first step is, therefore, to filter the data so we are left with only those bits of DNA that exhibit changes from DNA in normal cells and so we reduce the volume of data that we need to display graphically (imagine trying to display information for 20,000 genes, rather than say, 1000 in which there are changes). Instead of using numbers on a graph to indicate changes, many methods for displaying multidimensional use colour. Zero to nine can be represented by changing the depth of colour from one shade of green to another (Figure 4.20, colour version is located in the colour plate section). Equally, this colour scale could represent any relevant scale. Any numbers can be represented in this way as long as we include a colour scale bar as a key so that the reader knows which depth or shade of colour represents which number.

Figure 4.20   Numbers represented on a colour scale from light green to dark green. The colours could represent any range of numbers as long as a scale bar is clearly reported. (For a colour version, see the colour plate section.)

Changes can also be represented by using negative numbers for decreases and positive numbers for increases and more than one colour. Commonly, gene expression changes are represented as green for decreases (down-regulation), red for increases (up-regulation). No change might be represented as the transition between red and green, or by black. In Figure 4.21 (a colour version is located in the colour plate section), we have converted the numbers in Table 4.1 to a heat map, a plot representing numbers by intensity of colours. Although popular, green and red are indistinguishable to colour blind observers so some people use blue and orange, yellow and purple or yellow and red as alternatives. For gene expression data (the application for which heat maps are most commonly used in genomics), columns and rows represent samples and genes, so each square is the colour value of one gene in one sample. The difference in gene expression can vary greatly between genes and the range may be too wide to be distinguishable by shades of one or two colours; imagine trying to represent numbers ranging from 10 to 1,000,000, 200 would likely not be distinguishable from 2000. To circumvent this problem the values are often converted to a log scale (commonly $\log_2$ fold changes for data from microarrays, but $\log_{10}$ copy numbers could equally be plotted) which compresses the data permitting the colour gradation to be visualised. While heat maps may be regarded as the classic means of visualising genomic data, there are now many other types of colour map now in use and many computer programs for creating them easily, for example, CIRCOS software generates colour maps in a circular format (see www.circos.ca).

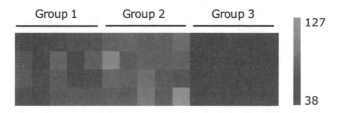

Figure 4.21 Heat map representing the data in Table 4.2. These colour graphs are commonly used to summarise genomic data such as gene expression changes. Green to red is very popular but may be difficult for colour-blind readers to distinguish differences so blue to orange or yellow to red are better. (For a colour version, see the colour plate section.)

4.8   Displaying proportions or percentages

4.8.1   The pie chart

Pie charts are commonly used to summarise percentages, proportions or information such as how people responded in a questionnaire (Figure 4.22). In the latter case, the chart is generated using the number of respondents but the labels could be response categories rather than numbers, for example, once, yearly, monthly, weekly (Figure 4.22b).

4.8.2   Tabulation

Graphs are not the only way in which data may be summarised. The effectiveness of communicating results by using tables is often underestimated. Tables are particularly useful for presenting data such as percentages or proportions

(a) (b)

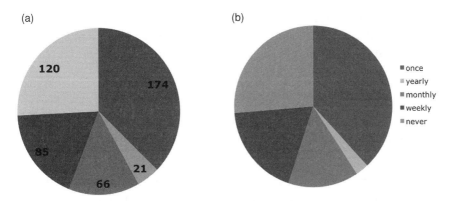

Figure 4.22 Pie chart of the same data plotted as (a) actual numerical values represented by the slices and (b) counts in categories represented by the slices.

with their errors. Percentage and proportion data should be used with discretion, because they can be misused to exaggerate a difference (a 100% increase is simply twice as much), but there are occasions where they are the most appropriate. For example, if we were adding a novel cytotoxic compound to a cancer cell line in tissue culture plates, because we wanted to know how effective it was at killing the cells, it would not make sense to report the absolute number of dead cells, as each well would have contained a different number of cells at the start of the experiment. It is better to express the number of dead cells as a percentage of the original number of cells in that well. The data may be clearly reported in a table with means and standard deviations of the replicate values (Table 4.4). Note that we have not used the word 'control' but rather specified the actual control treatments. It is a common mistake to simply state 'control' but this is not a useful descriptor. Controls can be negative or positive and there are many types of control, for example, in Table 4.4 there are two types of negative control. One to check that there is no effect of simply keeping cells in culture (culture medium alone) and the other to determine if it is the substance in which the compound is dissolved, the carrier that

Table 4.4 Percentage of viable cells determined by flow cytometry following treatment with a cytotoxic compound over 3 days. The compound was dissolved in 1% DMSO and treatments were reapplied daily. Data are mean percentage viable cells $\pm$ standard deviations, $n = 6$

Time (hours)	0	12	24	36	48	72
Culture medium	98.2 ± 0.5	96.4 ± 1.1	97.3 ± 2.0	94.3 ± 1.5	95.8 ± 2.3	96.0 ± 2.4
1% DMSO alone	97.5 ± 1.2	92.4 ± 1.1	91.2 ± 4.1	93.6 ± 3.5	90.5 ± 2.3	88.7 ± 1.4
Cytotoxic compound	96.8 ± 0.9	93.5 ± 2.1	89.4 ± 3.6	80.6 ± 4.5	71.5 ± 5.1	55.2 ± 5.2

produces the effect rather than the active compound itself. As well as descriptive statistics, the results of statistical tests can also be included in a table, for example, confidence intervals, the test statistic and probability.

4.9 Summary

- Raw data are conveniently summarised numerically using descriptive statistics such as a measure of central location and a measure of dispersion.

- The appropriate descriptive statistics should be selected depending on whether the data are symmetrically distributed or not.

- Descriptive statistics can be visualised graphically.

- The primary aim of summarising data is to provide a transparent and representative description of the original data.

- Graphs can illustrate group data or relationships between variables.

- Large data sets may be condensed and reported using colour as a surrogate for numbers.

References

Grafen, A. and Hails, R. (2002) *Modern Statistics for the Life Sciences*. New York: Oxford University Press.

Joiner, B.L. (1975) Living histograms. *International Statistical Review*, 43(3):339–340.

5
Statistical Power – '... Find out the Cause of this Effect'

Since statistical significance is so earnestly sought and devoutly wished for ... one would think that the a priori probability of its accomplishment would be routinely determined and well understood. Quite surprisingly, this is not the case.

Jacob Cohen

5.1 Aims

This is an ambitious chapter. In it we will take you through the interlinking issues of statistical power and effect size. In order to get to grips with some of these concepts it will be necessary to work through examples. The *p* value obtained in a statistical test can be misleading. One way of clarifying its use is through confidence intervals (CIs). Their use is becoming more popular in the presentation of results, and allows us to appreciate the size of the effect and the power of the statistical test. Understanding the concepts introduced in this chapter should significantly raise your statistical awareness and competency.

5.2 Power

Obtaining $p < .05$ in a statistical significance test is part of a well-established procedure (even ritual) in science necessary to persuade others that there **may** be something worthwhile in your findings. We can claim that there is

Starting Out in Statistics: An Introduction for Students of Human Health, Disease, and Psychology
First Edition. Patricia de Winter and Peter M. B. Cahusac.
© 2014 John Wiley & Sons, Ltd. Published 2014 by John Wiley & Sons, Ltd.
Companion Website: www.wiley.com/go/deWinter/startingstatistics

'statistical significance', that is, there is only a small probability that these results, or more extreme, could have arisen by chance **were the null hypothesis true**. Often attaining a small p value ensures that the research will get published. Now, if there really is an effect, and the null hypothesis is false, it would seem advisable to try to maximise one's chances of obtaining a statistically significant result. The probability of obtaining a statistically significant result, if there really is an effect present, is known as the *statistical power* of the test. The greater this probability is, the more powerful the test is. As the statistical power increases, the probability of a Type II error decreases. Recall that the Type II error is the probability of failing to reject a false null hypothesis (like the law court scenario of the guilty going free mentioned in Chapter 3, Table 3.2). Both the power and the probability of a Type II error are conditional on the null being false (i.e. the alternative hypothesis being true: that there is a real effect present). Using the court example, we must imagine that each of the accused who appear on trial is in fact guilty. Then the probability of them being released represents a Type II error, while the probability that they are convicted represents the power. If the null hypothesis happens to be true then power and Type II error do not exist, and cannot be calculated. Again in our court scenario, if all those charged were innocent then it makes no sense to talk about the probabilities of incorrectly releasing them or correctly convicting them.

So what factors would you think might increase statistical power? How might we improve your study to ensure that we get a statistically significant result when a real effect is present? Obviously if there is a big effect, like the height difference between men and women, then that would make it easier to reject the equal height null hypothesis. But what if we only have two men and two women? With so few in each sample, it is unlikely that we will find a statistically significant difference. It is certainly unlikely that we would convince others of such a difference with so little data. So increasing the number in the two samples will increase the power.

Another way to increase power is to reduce the variability of the data points. As long as this does not bias the collection of data to obtain a pre-specified objective, then this is valid procedure. One common way is to restrict one's data, typically on the grounds of age, gender, race, socioeconomic status, etc. – depending on the nature of the study. However, applying such restrictions to the data will limit how far you can generalise the results of your statistical test. In our height example, we could reasonably restrict age to 20–50 year olds and also to a particular race, let us say ethnic Orcadians. However, you must keep in mind that the restrictions that have been made in the selection of samples then apply to the inferences we make about the populations being studied. In our case, we cannot generalise outside of the 20–50 year old age group (very young children's heights differ little across gender). Also, our conclusions will only apply to the Orcadians population.

Table 5.1 Power and the probability of making a Type II error are conditional on the null hypothesis being false (shaded), that is, a real effect being present. If β is the probability of a Type II error, then the power is $1 - \beta$

	H_0 true	H_0 false
H_0 rejected	Type I error	Correct decision
	α	$1 - \beta$
H_0 not rejected	Correct decision	Type II error
	$1 - \alpha$	β

If we look at the decision table in Table 5.1 we can see that by increasing the statistical power we are reducing the probability of a Type II error, that is, β is reduced because power $= 1 - \alpha$. So we can think of attempts to increase the power of a statistical test as minimising the probability of a Type II error. Remember though, that the power of a test can only be considered or calculated when H_0 is false (see highlighted area in Table 5.1).

The *effect size* is an important concept related to statistical testing, but often forgotten in the excitement of obtaining a p value $< .05$. We mentioned above that where there is a big effect then it is easier to reject the null hypothesis. The effect size, or size of effect, is a calculated or estimated statistic (or parameter) which tells us how large an effect is – either actual or hoped for. It is a measure of the magnitude of the effect of an intervention/treatment relative to the null hypothesis value. A standardised form of effect size can be calculated from the data obtained in a completed study. For example, we could divide the difference between the sample mean and the null hypothesis mean by the sample standard deviation. Dividing by the standard deviation standardises the effect size, and is similar to the z value described below. Alternatively, the effect size can be estimated by proposing a particular value for a population parameter. If for example, before commencing a study, we were interested in a particular magnitude of difference from the null hypothesis. We would do this for one of two reasons. Either, because we would expect that value of the parameter from previous research, or, because this would be the minimum size of effect which would be of practical importance (see below in relation to Figure 5.7). In a similar way, we could divide the magnitude of the effect (difference between parameter and null hypothesis value) by a guestimate for the standard deviation. Depending on the motivation for the calculation, the proposed population parameter and the standard deviation guestimate could be obtained from the literature, from knowledge of the scientific area or from a pilot study. Such calculations would give us Cohen's d (Equation 5.1).

Equation 5.1 For calculation of Cohen's standardised effect size using population parameter estimates.

$$d = (\mu_1 - \mu_0) / \sigma \tag{5.1}$$

Effect size is used in calculations for sample size requirements before collecting data. Often we can suggest a priori an effect size for d. Typically, the following range of values is suggested: 0.2 to represent a small effect size, 0.5 a medium effect size and 0.8 a large effect size. Different measures for effect size are also used for other statistical analyses such as correlation (r, R^2) and ANOVA (partial η^2). Both R^2 and partial η^2 represent the proportion of overall variability explained by the effect of interest.

5.3   From doormats to aortic valves

As a slight diversion we can examine briefly how a decision procedure can be used in quality control. In manufacturing goods, for example, decisions need to be made with respect to setting a threshold at which to reject poor quality items. Let us imagine a simple decision process used to reject doormats on a production conveyor belt. The factors pertinent to doormat rejection or acceptance can be summarised as the quality (good or bad) and the decision to pass or fail (Table 5.2). As long as the mat is about the right size, colour and texture, it would be functional and sellable. The manufacturer does not want to reject too many mats that are of good quality since that reduces profits, so the probability α will be minimised. The value of α is the probability of discarding good (sellable) mats, and is known as the *producer's risk*. If the mats are good (left column in Table 5.2) then the remaining probability of passing good quality mats is $1 - \alpha$. The two probabilities for the two possible decisions here must add up to 1. The other type of error consists of poor quality mats which are passed as acceptable but really are not, hence the consumer suffers. This error is shown across the diagonal, and is the *consumer's risk*, symbolised as β (Table 5.2). If these are cheap mats then the consumer may cut their losses, dispose of the mat and obtain a replacement. Of course, if consumers

Table 5.2 Decision table for the production of doormats. Relatively speaking, quality control is not crucial, the main concern is to minimise producer's risk, aka Type I error or α. The manufacturer might want to minimise this to some small probability, such as 1 in a thousand. If the mats are good we would pass a proportion corresponding to $1 - \alpha$ (999 / 1000 good mats). Bad mats that are not rejected become the consumer's risk, aka Type II error or β

	True quality of doormat	
	Good	Bad
Fail	α (producer's risk) ✗	$(1 - \beta)$ Power ✓
Pass	$(1 - \alpha)$ ✓	β (consumer's risk) ✗

Table 5.3 If consumer satisfaction was critical (e.g. patient survival, subject to legal action and compensation), then the consumer's/patient's risk would be chosen to be α, as shown here (table columns reversed from previous Figure). The probability of α might then be very small, probably less than 1 in a million

		True state of aortic valve	
		Bad	Good
Decision	Pass	α (patient's risk) ✗	$(1 - \beta)$ Power ✓
	Fail	$(1 - \alpha)$ ✓	β (producer's risk) ✗

are unhappy with a product for which they have paid good money they will not buy again, and they may also tell all their friends and Facebook followers about their 'atrocious' experience, which may in turn lead to lower future sales for the manufacturer. The 'power' here would be the probability of the quality control process failing to identify those mats that are of poor quality, and calculated as $1 - \beta$. Again the two mutually exclusive probabilities here (right column of Table 5.2) must add up to 1.

In contrast, we can consider a quite different scenario in medical manufacturing. In the production of aortic valves for cardiac surgery, the consumer's risk becomes of paramount importance. Any manufacturing defect that leads to injury or death of the patient (= consumer) would have severe legal and financial repercussions for the manufacturer. In this case the consumer's risk must be minimised. This is illustrated in Table 5.3, which shows a re-arranged decision table where **Bad** and **Good** have been swapped round. Having very tight criteria for the acceptance of aortic valves would also mean rejecting many perfectly good valves. The probabilities of α and β are inversely related (as α shrinks, β increases). Pulling a figure out of the air, we will suppose a large producer's risk for β such as .6 or 60%. This risk (cost/wastage) is covered by the very high cost of the final product. In this scenario, the 'power' of the procedure consists of the probability of correctly deciding to pass a good valve. Given the relatively large producer's risk, this probability might only be .4 or 40% $(1 - \beta)$.

These examples illustrate how the relative importance of the two types of errors, α and β, makes a difference to our decision-making framework. One could imagine a research scenario where the probability of a Type I error should be minimised. For example, for an intervention that is very expensive or associated with adverse side-effects, deciding to administer it might cause more harm than good, if it is actually ineffective. By contrast, if a treatment is very cheap and has no adverse effects then making a Type I error, α, would be less important ('better safe than sorry' approach). Here, we may be more concerned about minimising a Type II error, β, where a potentially useful treatment might be missed.

5.4   More on the normal distribution

Before we can proceed with an example of the calculation of statistical power, we need to recall the previous chapter where the ever important normal distribution was introduced. Examples were given of variable distributions that were, or were not, normally distributed. The normal distribution is a mathematical ideal to which real data only approximate. Nevertheless, the approximation is good enough for many situations that we encounter in statistics. In the next chapter, for example, you will see that its existence is assumed for many types of statistical tests. Since the normal distribution is mathematically defined we can use it to obtain precise areas under the normal distribution curve. Why on earth would we want to know areas under the curve? If we look at Figure 5.1 we can see the familiar bell-shaped curve. This might represent (approximately) the population of measurements taken from an animal. Imagine that the population of the threatened diminutive subspecies of the Santa Cruz Island island fox (*Urocyon littoralis santacruzae*) has been exhaustively studied by a zoologist who measures tail lengths of every living adult male. If the whole population of adult male island foxes was measured (a census), then we can obtain the population mean and standard deviation. Say the tail lengths' population mean μ is 6.6 inches and its standard deviation σ is 0.8 inch, and that the distribution of tail lengths closely followed the

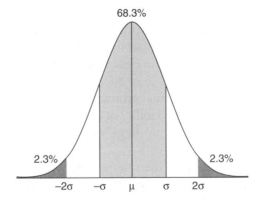

Figure 5.1 The areas under the normal curve are precisely defined according to the number of standard deviations from the mean. Hence, the area within 1 standard deviation of the mean contains approximately 68.3% of the total area (middle light grey area). Each tail area that is 2 standard deviations from the mean contains approximately 2.3% of the total area (dark grey area).

normal distribution. Most observations are close to the mean. The frequency of observations diminishes on either side as we move further away from the mean and into the tails of the distribution. The filled areas illustrate specific areas in relation to one and two standard deviations from the mean. The larger of these (in light grey), is the area within (± 1) standard deviation of the mean. More than two thirds, approximately 68.3%, of observations are in this area. There are far fewer observations (e.g. fewer very short and very long tails) more than two standard deviations on either side of the mean. Each of the dark grey shaded areas contains only about 2.3% of the total population. So, less than 5% of the tail lengths are beyond 2 standard deviations from the mean. Obviously the whole area under the curve represents all tail lengths, 100%. Because the relationship between standard deviations from the mean (horizontal axis) and frequency (vertical axis) can be calculated (see below) it means that we can find out how much of the population lies between any two specified points on the standard deviation axis.

To avoid using a specific value for the mean and standard deviation of a set of data we *standardise* the distribution, to give us the *standard normal distribution* (Figure 5.2). We explain how this is done below, but first note that the figure is identical with the previous one, except for the horizontal axis. The standard normal distribution is centred at 0. On either side of 0 the numbers represent distance from 0 in terms of standard deviations. The horizontal axis is now represented by the letter z. If we know the standard deviation of the population (σ) and the mean, then we can convert any values from that distribution into z values using Equation 5.2:

Equation 5.2   Conversion of observations into z values.

$$z = (\text{Observation} - \text{mean}) / \sigma \tag{5.2}$$

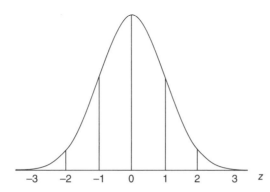

Figure 5.2   The standard normal distribution is exactly the same as the normal distribution except that the mean or centre of the distribution is at 0, and the units on the horizontal axis refer to standard deviations and are represented by z.

We can see from this equation and Figures 5.1 and 5.2 that all z represents is the number of standard deviations away from the mean a particular value is. If it is negative it is below the mean, if positive it is above the mean.

The conversion of measurements to a z value allows us to calculate probabilities or proportions. Also, given probabilities or proportions we can calculate z values and hence values of tail lengths. For example, we could rather precisely estimate the proportion of island foxes that have tail lengths greater than 9 inches.

Step 1. Calculate how far 9 is from 6.6 in terms of standard deviations, that is, the z value:

$$z = (9 - 6.6) / 0.8$$
$$= 3.000$$

Step 2. Look up the z value in Table 5.4 to find the corresponding tail area. The relevant row for this z value in the table is:

z	0	0.01	0.02	0.03	0.04	0.05	0.06	0.07	0.08	0.09
3.0	0.0013	0.0013	0.0013	0.0012	0.0012	0.0011	0.0011	0.0011	0.0010	0.0010

The shaded value of **.0013 (or 0.13%)** represents the proportion of island foxes that have tails lengths exceeding 9 inches. This is the same as saying that the probability of a randomly selected fox from the population having a tail exceeding 9 inches is .0013.

We can do more involved calculations like: what proportion of island foxes have tail lengths between 8 and 9 inches?

Step 1. We need to convert both of these measurements in inches to z values. We have done that already for 9 inches. Let's do the same for 8 inches:

$$z = (8 - 6.6) / 0.8$$
$$= 1.750$$

Step 2. Look this value up in Table 5.4, the relevant row being:

z	0	0.01	0.02	0.03	0.04	0.05	0.06	0.07	0.08	0.09
1.7	0.0446	0.0436	0.0427	0.0418	0.0409	0.0401	0.0392	0.0384	0.0375	0.0367

The shaded value is .0401.

Table 5.4   The z table showing tail areas under the normal curve. The z value is represented to one decimal place by the values in the left margin and a further decimal place by the values along the top margin of the table. For example, the z value corresponding to 5% significance for two tails (.025% in each tail) we find in the body of the table the value 0.025, and then read across to the left (1.9) and top (0.06) to obtain the z value of 1.96 (shaded)

z	0	0.01	0.02	0.03	0.04	0.05	0.06	0.07	0.08	0.09
0.0	0.5000	0.4960	0.4920	0.4880	0.4840	0.4801	0.4761	0.4721	0.4681	0.4641
0.1	0.4602	0.4562	0.4522	0.4483	0.4443	0.4404	0.4364	0.4325	0.4286	0.4247
0.2	0.4207	0.4168	0.4129	0.4090	0.4052	0.4013	0.3974	0.3936	0.3897	0.3859
0.3	0.3821	0.3783	0.3745	0.3707	0.3669	0.3632	0.3594	0.3557	0.3520	0.3483
0.4	0.3446	0.3409	0.3372	0.3336	0.3300	0.3264	0.3228	0.3192	0.3156	0.3121
0.5	0.3085	0.3050	0.3015	0.2981	0.2946	0.2912	0.2877	0.2843	0.2810	0.2776
0.6	0.2743	0.2709	0.2676	0.2643	0.2611	0.2578	0.2546	0.2514	0.2483	0.2451
0.7	0.2420	0.2389	0.2358	0.2327	0.2296	0.2266	0.2236	0.2206	0.2177	0.2148
0.8	0.2119	0.2090	0.2061	0.2033	0.2005	0.1977	0.1949	0.1922	0.1894	0.1867
0.9	0.1841	0.1814	0.1788	0.1762	0.1736	0.1711	0.1685	0.1660	0.1635	0.1611
1	0.1587	0.1562	0.1539	0.1515	0.1492	0.1469	0.1446	0.1423	0.1401	0.1379
1.1	0.1357	0.1335	0.1314	0.1292	0.1271	0.1251	0.1230	0.1210	0.1190	0.1170
1.2	0.1151	0.1131	0.1112	0.1093	0.1075	0.1056	0.1038	0.1020	0.1003	0.0985
1.3	0.0968	0.0951	0.0934	0.0918	0.0901	0.0885	0.0869	0.0853	0.0838	0.0823
1.4	0.0808	0.0793	0.0778	0.0764	0.0749	0.0735	0.0721	0.0708	0.0694	0.0681
1.5	0.0668	0.0655	0.0643	0.0630	0.0618	0.0606	0.0594	0.0582	0.0571	0.0559
1.6	0.0548	0.0537	0.0526	0.0516	0.0505	0.0495	0.0485	0.0475	0.0465	0.0455
1.7	0.0446	0.0436	0.0427	0.0418	0.0409	0.0401	0.0392	0.0384	0.0375	0.0367
1.8	0.0359	0.0351	0.0344	0.0336	0.0329	0.0322	0.0314	0.0307	0.0301	0.0294
1.9	0.0287	0.0281	0.0274	0.0268	0.0262	0.0256	0.0250	0.0244	0.0239	0.0233
2	0.0228	0.0222	0.0217	0.0212	0.0207	0.0202	0.0197	0.0192	0.0188	0.0183
2.1	0.0179	0.0174	0.0170	0.0166	0.0162	0.0158	0.0154	0.0150	0.0146	0.0143
2.2	0.0139	0.0136	0.0132	0.0129	0.0125	0.0122	0.0119	0.0116	0.0113	0.0110
2.3	0.0107	0.0104	0.0102	0.0099	0.0096	0.0094	0.0091	0.0089	0.0087	0.0084
2.4	0.0082	0.0080	0.0078	0.0075	0.0073	0.0071	0.0069	0.0068	0.0066	0.0064
2.5	0.0062	0.0060	0.0059	0.0057	0.0055	0.0054	0.0052	0.0051	0.0049	0.0048
2.6	0.0047	0.0045	0.0044	0.0043	0.0041	0.0040	0.0039	0.0038	0.0037	0.0036
2.7	0.0035	0.0034	0.0033	0.0032	0.0031	0.0030	0.0029	0.0028	0.0027	0.0026
2.8	0.0026	0.0025	0.0024	0.0023	0.0023	0.0022	0.0021	0.0021	0.0020	0.0019
2.9	0.0019	0.0018	0.0018	0.0017	0.0016	0.0016	0.0015	0.0015	0.0014	0.0014
3.0	0.0013	0.0013	0.0013	0.0012	0.0012	0.0011	0.0011	0.0011	0.0010	0.0010
3.1	0.0010	0.0009	0.0009	0.0009	0.0008	0.0008	0.0008	0.0008	0.0007	0.0007
3.2	0.0007	0.0007	0.0006	0.0006	0.0006	0.0006	0.0006	0.0005	0.0005	0.0005
3.3	0.0005	0.0005	0.0005	0.0004	0.0004	0.0004	0.0004	0.0004	0.0004	0.0003
3.4	0.0003	0.0003	0.0003	0.0003	0.0003	0.0003	0.0003	0.0003	0.0003	0.0002
3.5	0.0002	0.0002	0.0002	0.0002	0.0002	0.0002	0.0002	0.0002	0.0002	0.0002
3.6	0.0002	0.0002	0.0001	0.0001	0.0001	0.0001	0.0001	0.0001	0.0001	0.0001
3.7	0.0001	0.0001	0.0001	0.0001	0.0001	0.0001	0.0001	0.0001	0.0001	0.0001
3.8	0.0001	0.0001	0.0001	0.0001	0.0001	0.0001	0.0001	0.0001	0.0001	0.0001
3.9	0.0000	0.0000	0.0000	0.0000	0.0000	0.0000	0.0000	0.0000	0.0000	0.0000
4	0.0000	0.0000	0.0000	0.0000	0.0000	0.0000	0.0000	0.0000	0.0000	0.0000

Step 3. The value we obtained in Step 1 represents the proportion of island foxes with tail lengths exceeding 8 inches. We have the proportion for exceeding 9 inches, therefore the proportion lying between 8 and 9 inches will be the proportion 8 inches minus the proportion for 9 inches:

$$.0401 - .0013 = .0388$$

Rounding this to three decimal places we get **.039 (or 3.9%)** which is the proportion of island foxes with tail lengths between 8 and 9 inches long.

Or we could calculate the proportion of foxes with tail lengths less than 6 inches long.

Step 1. Calculate the z value for the difference between 6 and 6.6:

$$z = (6 - 6.6) / 0.8$$
$$= -0.750$$

Step 2. Look up the tail area for -0.75. This is a negative value which we obtained because our value of interest is less than the mean value. We ignore the negative sign because by symmetry the areas are the same for positive values. This should give us **.2266 (or 22.7%)** as the proportion of island foxes with tails shorter than 6 inches.

We can also calculate the length of tail which is exceeded by the longest 25% of island fox tails.

Step 1. As 25% here is area in the normal distribution tail, we have to find this value (.25) in the body of Table 5.4. Here is the relevant row from the table:

z	0	0.01	0.02	0.03	0.04	0.05	0.06	0.07	0.08	0.09
0.6	0.2743	0.2709	0.2676	0.2643	0.2611	0.2578	0.2546	0.2514	0.2483	0.2451

The exact value for .25 does not appear in the table and the two closest values are shaded. These represent a z value of between 0.67 and 0.68, so we will interpolate roughly to give us a z value of 0.675.

Step 2. This z value represents the number of standard deviations **above** the population mean (it would be below the mean if we were considering the shortest tail lengths). Since σ is 0.8 we multiply this by our z value, and add it to μ:

$$= 6.6 + (0.8 * 0.675)$$
$$= 7.14 \text{ inches}$$

Hence, the length of tail above which the longest 25% of tails lie is **7.14 inches**. This is the same as the value for the third *quartile* (ordering the data from smallest to largest, this value is three quarters of the way along). It also means that 75% of tail lengths lie below this value, because obviously the whole area under the normal curve adds up to 100% (or probability 1).

Still not clear why we need to know any of this? OK, the areas under the normal distribution curve represent probabilities, and statistical tests give us probabilities (p values). You will see in the next chapter, when we talk about the distribution of means (which is also tend to be normally distributed), we can answer questions about means such as 'What is the probability of obtaining a randomly selected sample mean more extreme than this, assuming the null hypothesis is true'?

5.4.1 The central limit theorem

We need a brief, but important, diversion from z values to try to understand why the normal distribution is so fundamental to much statistical theory. We have largely discussed individual observations above, for example, island fox tail lengths – how they are distributed normally. However, in research studies we are rarely interested in individual values. Typically, we collect a sample of data and calculate its mean. So we are interested in how the **means are distributed**. That is, if we randomly selected many samples of island foxes from a large population, calculated the mean tail length for each sample and plotted all the means in a histogram, what would that histogram look like? It will be no surprise to learn that if the individual observations (say the tail lengths) are normally distributed, then the distribution of means will also be normally distributed (Figure 5.3). The distribution of means is known as the *sampling distribution of the mean*, or just the *sampling distribution*. What may surprise you is that if the individual values of a population are **not** normally distributed, then the sampling distribution will become normally distributed if the sample size n is large. Uniform (Figure 5.4) and skewed (Figure 5.5) distributions, even binary distributions (e.g. 0, 1), will produce sampling distributions, which are approximately normally distributed. The variance of the sampling distribution will be σ^2 / n, meaning it is inversely related to the sample size. The larger the sample the smaller the dispersion of the sampling distribution, that is, the sampling distribution gets narrower. The standard deviation

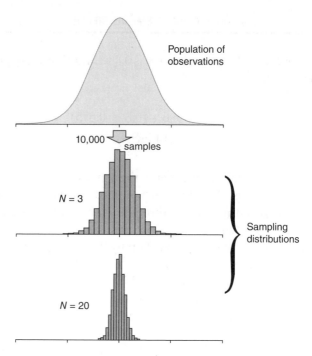

Figure 5.3 The sampling distributions obtained from a normal distribution are normal. The top histogram represents the distribution of the population (light grey). The two lower distributions (dark grey) represent sampling distributions of 1000 random sample means taken from the population. The upper of the two sampling distributions has a sample size of three ($N = 3$), while the lower sampling distribution has a sample size of twenty ($N = 20$).

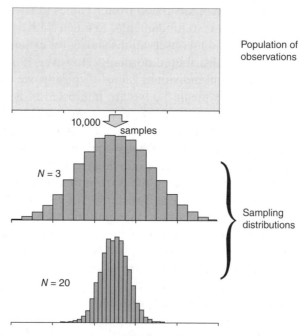

Figure 5.4 The sampling distributions obtained from a uniform distribution becomes normal, and the larger the sample size the more normal the sampling distribution becomes. Details as in Figure 5.3.

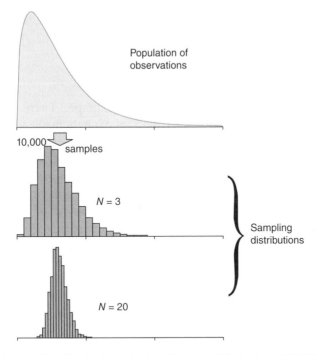

Figure 5.5   The sampling distributions obtained from a positively skewed distribution becomes normal, and the larger the sample size the more normal the sampling distribution becomes. Details as in Figure 5.3.

of the sampling distribution will be the square root of the variance, $\sigma / \sqrt{n}$, also known as the *standard error of the mean* (SEM). As sample size increases, the sampling distribution will tend to be ever closer to a normal distribution, and the mean of the distribution will approach the population mean μ. The *central limit theorem* specifies important properties of the sampling distribution. Since we usually take a sample, rather than individual observations, the central limit theorem tells us how such a sample is likely to behave, and moreover, that it will be normally distributed when a larger sample is used. This is why the normal distribution is so important in statistics.

Let's return to z values and how we can understand statistical power.

5.5   How is power useful?

Recall from our doormats and aortic valves examples that power was mentioned. This represented making the right decision (to fail bad mats or to pass good valves). In statistical testing this is known as statistical power and represents the probability of correctly rejecting the null hypothesis, when the null hypothesis is false (see Table 5.1). This allows us to determine how likely our study, with a given sample size, would get a statistically significant result if there really was an effect present.

5.5.1   Calculating the power

To illustrate this we can work through a simple example. There are computer games which claim to be able to 'brain train' – playing them allegedly increases your IQ. If a study had been done using 25 participants and we assume that they have a mean IQ of 100 ($\mu = 100$). Let us also assume that an increase in their mean IQ from 100 to 105 would be practically important, anything less would be too trivial for us to consider the intervention (playing the computer game) to be worthwhile. **We want to calculate the power of the test given a sample size of 25 participants**. The population standard deviation σ for the IQ test used is 10^1, therefore our standard error for the mean SEM = 10 / $\sqrt{n}$ = 10 / $\sqrt{25} = 2$. Our effect size is obtained from the difference between 105 and 100 (population mean), divided by σ, giving us 5 / 10 = 0.5 (recall from above, this is a medium-sized effect). As usual we will use $\alpha = .05$. We will also use what is called a *two-tailed test*. A two-tailed test allows for the possibility of the effect to be positive (an increase in the IQ) or negative (a decrease in the IQ). Unless you have very good reasons (and without looking contrived), you should always use two-tailed statistical tests.

We next need to find the *critical z* value for the statistical significance level α. The critical z value corresponds to the z value for the areas in the tail of the normal distribution where we would reject the null hypothesis using a significance level of 5% (α). This can be obtained from the z table given in Table 5.4. Since we are using a two-tailed test, and 5% is our significance level, we divide the 5% by 2 to give us 2.5% in each tail area. Alternatively, z can be directly calculated in **Excel** by typing into one of the cells this formula:

$$= \text{NORM.S.INV}(0.975) \text{ in version 2010, or} = \text{NORMSINV}(0.975)$$
$$\text{in version 2007}$$

.975 is the probability obtained by 1 − (.05 / 2). The .05 is divided by 2 because we are doing a two-tailed test. So it is giving us the number of standard errors from the mean which includes 97.5% of the curve (the two tails contain .025, which together add up to .05). Whichever way you use, you will get a z of 1.96.

---

[1] IQ tests are devised so that they have a specified mean (100) and standard deviation for a given population. Some IQ tests are devised to have a standard deviation of 15 instead of 10.

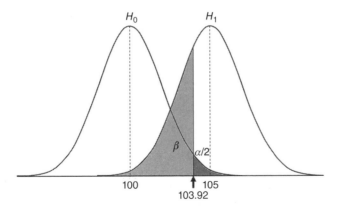

Figure 5.6 The sampling distributions according to whether the null hypothesis is true (left curve) or not (right curve). The critical region for rejection of H_0 is shown dark grey, and it has an area of α / 2 = .05 / 2 = .025. The arrow indicates the point at which the critical region begins, and was calculated by 1.96 * SEM = 103.92. The light grey area labelled β represents the probability of a Type II error. The area to the right of the light grey area (in the right curve) represents the power of the study $(1 - \beta)$.

In relation to the sampling distribution of the IQ means, the critical value is calculated by:

$$\mu + (z * \text{SEM})$$
$$100 + (1.96 * 2) = 103.92$$

Recall from Chapter 3, the symbols we use for the null (H_0) and the alternative (H_1) hypotheses. The position of this mean is shown by the arrow below the H_0 curve in Figure 5.6. If a sample mean is obtained in the study that is equal or greater than this, then H_0 is rejected. The area under H_0 true (left) curve to the right of this point, dark grey shading, is therefore 2.5% (α / 2). The other area, shaded in light grey to the left of this point but under the H_1 is true (right) curve, represents the probability of making a Type II error, denoted β. Assuming H_1 is true, and then if a sample mean obtained by our study is less than 103.92 then we would not reject H_0, hence we would be making a Type II error. The remaining area under the H_1 true curve is $1 - \beta$, and is the statistical power. To calculate this we need to find the number of SEMs away from the mean if H_1 is true. Remember, our smallest value to be of interest was an IQ of 105. In this analysis we are assuming the true value is 105, but in the study we may get values a little less than that due to sampling variability. It is possible then that when we carry out our study we could get a mean IQ of less than 105 and yet this would be statistically significant (if >103.92). Yet, we have specified 105 to be the minimum value of interest. This could

well be due to sampling variability from the true value of 105, or it may mean
that the true value was really less than 105. Since the researcher has specified
105 as the minimum they may need to discount the finding as statistically but
not practically significant.

Now we must calculate the number of standard errors from 105 to the arrow
(103.92):

$$z = (103.92 - 105) / 2$$
$$= -0.54$$

Notice we have a negative z value here. It is negative because the critical
value is to the left of the H_1 mean. The standard normal (z) distribution is
symmetrical (a negative z value table would be identical to the positive one
that we use), so we merely look up the value for 0.54. From the z table find
0.54 in the z column and read across to obtain the corresponding **Tail** area,
which will be 0.2946. This is β. Power is $1 - \beta$, which then gives us $1 - 0.2946 =$
0.7054 or 70.54%.

This result tells us that if the real effect of the computer game was to raise
the IQ by 5 points, and we use a sample of 25 people, then we have at least a
70% probability of getting a statistically significant result ($p < .05$). Well, this
is better than tossing a coin, but it is not great. Many years ago Jacob Cohen,
who was quoted at the start of the chapter, reviewed the power of published
studies in psychology and found the mean power employed was about 50%,
that is, researchers were as good as tossing a coin! You can imagine how many
fruitless studies were performed with little hope of achieving statistical significance. Most studies now aim to have a power of at least 80%. Why would
we have bothered to do the above calculations? Typically they would be done
post hoc (which means after the event, i.e. once the study had been completed)
to see what the sensitivity (power) was. For example, if the study had failed
to be statistically significant it could be pointed out that the sensitivity was on
the low side. How could we increase the sensitivity? – Answer: by increasing
our sample size.

5.5.2   Calculating the sample size

Let us say that we wanted to have a good chance of correctly rejecting H_0. A
value of 90% for power would then be appropriate. We now need to calculate
the sample size to be used in this proposed study. In Figure 5.6 we now know
that the light grey shaded area representing β would be .1 (1 − .9). We look up
this value in the **Tail** column (nearest value) in the z table to find the corresponding z value of 1.28. This means that 105 is 1.28 standard errors away from
the point on the axis representing the critical value (i.e. 105 − (1.28 ∗ SEM).
We know, again from looking at Figure 5.6, that this is also the same position

as $100 + (1.96 * \text{SEM})$. So these two expressions are the same, referring to the same point, hence we get:

$$105 - (1.28 * \text{SEM}) = 100 + (1.96 * \text{SEM})$$

Rearranging (subtracting 100 from both sides, and adding $1.28 * \text{SEM}$ to both sides) we get:

$$105 - 100 = (1.28 + 1.96) * \text{SEM}$$
$$5 = 3.24 * \text{SEM}$$

since $\text{SEM} = \sigma / \sqrt{n} = 10 / \sqrt{n}$

$5 = 3.24 * 10 / \sqrt{n}$ and rearranging (both sides: multiply by $\sqrt{n}$ and divide by 5) to give $\sqrt{n} = 32.4 / 5$, and hence $n = 41.9904$. So we will need 42 participants.

This means to increase our power to 90% from the earlier study scenario of about 70% (that used 25 participants), we will need 42 participants. This is normally how studies increase their power, by increasing the sample size. The relationship between power and sample size for this example is shown in Table 5.5. As noted above, we can also increase the power by reducing variability which will reduce the standard error. Also too, the larger the effect of the computer game playing on IQ the greater will be the power. Conversely, if we need to detect small effects we will need a large sample.

Hopefully the calculations above will give you a better understanding about what power is. This was done using the simplest example of a statistical test, the one sample test where we knew the population standard deviation. Typically we need to use more complicated statistical tests. For these, it is

Table 5.5 Relationship between power and sample size for our brain training example (assuming effect size of 0.5, $\sigma = 10$). The power with only eight participants is only 30%. For 99%, nine times the number of participants is required

Power (in %)	Sample size
99	73
95	52
90	42
80	31
70	25
60	20
50	15
40	12
30	8

recommended that you use a free downloadable software package such as G*Power.

5.6   The problem with *p* values

Actually there are a number of problems with *p* values, but hopefully we will now alert you to two of them (Table 5.6). In the table we have statistical significance (obtained or not obtained) as rows, and effect size (large or small) as columns. The top left and bottom right cells are fine using *p* values in statistical testing. In the top left we have a small *p* value $< .05$ together with a large effect size. This is good because we will be reporting an effect as statistically significant and it's a big effect. In the bottom right we have a large *p* value and a small effect size, meaning that we will not report this as statistically significant – which is fine because the effect size is anyway small. The problem cells are in the other diagonal. At bottom left we have our first problem, a large effect size (e.g. a big difference in means) but a large *p* value, meaning that we cannot report this as statistically significant. This suggests that the power of our test is too low, and we should have used a larger sample size, for example. The remaining cell at top right is the second problem; because we have a small *p* value but also a small effect size. This means that many journals would be happy to publish our work (and we may be happy too!) but since the effect size is small it might mean that the observed finding has negligible scientific/practical/clinical importance. This is one reason why science journals are unnecessarily cluttered with statistically significant findings which are of little or no practical importance. People are easily misled into thinking that statistical significance means practical importance.

Another problem with *p* values is that if the null hypothesis is true (there really is no difference/effect), then the probability of obtaining a statistically result $p < .05$ is unaffected by sample size. In contrast, if the null hypothesis is

Table 5.6   Problems are encountered using *p* values when they are large and the effect size is large (bottom left quadrant), and when *p* values are small and the effect size is small (top right quadrant). The first leads to possible failure to detect an important effect, while the latter detects trivial effects. Neither of these issues can be identified by the reported *p* value alone

		Effect size	
		Large	Small
Statistical significance	small *p* value <0.05	No problem	Mistaking statistical significance for scientific importance
	large *p* value >0.05	Failure to detect a scientifically important effect	No problem

Source: Adapted from Rosenthal *et al.* (2000).

false, then a statistically significant outcome can always be obtained if a sufficiently large sample is used. If a non-significant result is required (which is sometimes desired in research), then a smaller sample can be used to ensure that $p > .05$ (Surely, no one would do that?!) Researchers can misuse the fixed 'objective' significance level of .05 by subjectively selecting different sample sizes depending on which outcome is required. One way to avoid these issues is to use Bayes factor (as described in Chapter 3), but in addition taking into account sample size and effect size. In this way, the likelihood ratio (Chapter 3) will be driven towards the null hypothesis if it is more plausible than a specified alternative hypothesis value, and vice versa. This approach is convincingly described by Rouder *et al.* (2009) which has a website and R statistical software to carry out calculations.

5.7 Confidence intervals and power

When we estimate a population parameter, like a mean or correlation, we can calculate a CI around that estimate. Typically we would use a 95% CI to correspond to the commonly used 5% significance level. The 95% CI has limits on either side of the parameter estimate. What does this interval mean? Well, it is like a significance test but better. It can be used as a significance test because any value which lies outside of the interval can be rejected at the 5% significance level. Selecting a sample from a population allows us to calculate the sample mean as an estimate for the population mean (for example) and the 95% CI. The interval is the sample mean $\pm (t_{\alpha/2} * \text{SEM})$. The t here is similar to the z we used above, but we use it instead of z if we do not know the population standard deviation σ and must estimate it with the sample standard deviation, s. Each sample we take from the population will quite likely have a different mean (due to sampling variability) and also different variance; and hence the standard deviation and SEM will be different each time. Some CIs will be wide (large standard deviations) and others narrow (small standard deviations). If we were to calculate the 95% CI around the sample mean for each of many samples randomly selected from a population (with replacement), then in the limit – that is after innumerable samples, 95% of such samples would contain the true population mean. This is the frequentist definition. We can then think of any particular interval as having a 95% chance of containing the true population mean. The CI gives us a position or point estimate (in this case the mean) and also the limits of uncertainty around that point estimate. The width of the interval is a measure of precision for the population mean. If it is relatively wide this suggests poor precision (this happens with a small sample size). Conversely, a narrow interval suggests high precision. If we wanted to be more certain we could calculate a 99% interval (would it be wider or narrower than the 95% interval? – Answer: wider).

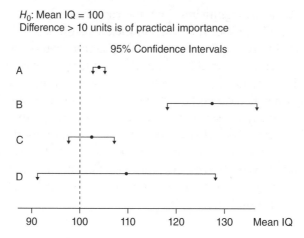

H_0: Mean IQ = 100
Difference > 10 units is of practical importance

Figure 5.7 Four study scenarios to illustrate the importance of confidence intervals. The mean IQ if the null hypothesis was true is 100 (dashed line). For an effect to be of practical importance it must exceed 10 units difference from 100. Scenarios A and B are both statistically significant with the same z and p values, but only B is of practical importance as the lower limit of the CI exceeds 110. In contrast, C and D are both not statistically significant, have the same z and p values, but only D is of interest as the CI includes values above 110. This suggests that in D there is insufficient data to claim evidence for an absence of effect. Adapted from Reichardt and Gollob (1997).

The relevance of CIs to power and effect size is illustrated in Figure 5.7. It shows 95% CIs from four study scenarios concerned with determining whether consumption of omega-3 supplements over a 6-month period had an effect on IQ. (You will notice that IQ is used a lot: that's because it is one of the few examples where we conveniently know the population mean and standard deviation.) If we assume that the mean IQ in each of the studies prior to omega-3 is 100, then we can compare the 95% CI after omega-3 with that value. So we are comparing the position of each CI with the dashed line representing no effect (the null). At the centre of each CI is the sample mean. Let us suppose that in order for an effect of the intervention to be considered practically important it must cause a greater than 10 unit difference (any value could be chosen and this will depend on the research area). So if omega-3 improves IQ we would be looking for differences that are statistically significantly greater than 110. The top two scenarios look very different from each other. A is close to the null effect with a narrow CI, while B is further away with a much wider CI. However the calculated statistics for each are identical (both $z = 6$), and hence the p values are identical too (both $p < .000001$). While A is statistically significant we can discount its importance because its effect size is negligible, being less than 110 (it is also significantly less than 110). Scenario B is different because it shows that the statistically significant effect is also of practical importance since the extent of the CI lies above 110.

Scenario B clearly indicates that omega-3 supplements are worthwhile, massively boosting IQ. If we look at the bottom pair of scenarios, C and D, we see that both of them are not statistically significant because their CIs include the null value of 100. Both their statistics are identical ($z = 1, p = .32$), but their mean positions and CI widths are different. C is narrow and the upper limit of its CI does not include 110. This suggests that if there is an effect of omega-3 in this study, it is negligible and is significantly different from 110. It is also likely that sufficient data were collected since the CI is relatively narrow. In scenario D the very wide CI spans the null value but also includes values of practical importance, almost up to 130. This suggests that we have insufficient data to make any reasonable decision. The study should be repeated with a larger sample. The required sample size could be calculated as we did earlier.

5.8 When to stop collecting data

In many studies data are collected sequentially, for example, in clinical studies where patients who are encountered with a particular disorder are assigned to a control or intervention group. Data may be collected over a number of months or years, and the question might be – when do we do our significance test to see if the intervention does anything? This issue may not arise if we have done our sample size calculations as described above, in which case we wait for the requisite number of patients to be enrolled in the trial. Otherwise we could consider carrying out our statistical test every time we add a few more patients – no, we cannot do this because the more times we test the more likely we are to obtain a statistically significant result, even if the null hypothesis is true (i.e. we increase our chances of making a Type I error). One elegant solution is to use a CI as a 'stopping rule' (Armitage *et al.*, 2001, p. 615). To pursue our example of IQs, we could consider an increase of 5 units or lower as negligible, but an increase of 10 units or above as important. What we then do is collect data until the 95% CI is less than 5 units **wide**. Initially with few patients the 95% CI will be very wide. As we increase our sample the CI will continue to shrink. Once we have achieved a CI less than 5 units wide we would check to see where the interval was relative to 105 and 110. In the example shown in Figure 5.8, our 95% CI excludes the trivial effect of 105 and includes the important value of 110, thus indicating that we have sufficient data to claim that the intervention is significantly better than a trivial effect. Our calculation for this procedure, if we knew σ (if not we could use t rather than z) and we were aiming for a 95% CI less than 5 units wide, would be to continue collecting data with N increasing until:

$$\left(\left(1.96 * \frac{\sigma}{\sqrt{N}} \right) + \bar{x} \right) - \left(\left(1.96 * \frac{\sigma}{\sqrt{N}} \right) - \bar{x} \right) < 5.00$$

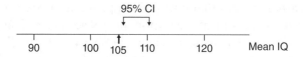

Figure 5.8 Illustrating the use of confidence intervals in a 'stopping rule', such that the CI narrows as the sample size increases. When the CI width falls below a pre-defined value, then we can check to see where the interval is relative to a negligible effect of 105 and an important effect of 110. In the scenario shown the CI spans 110 and excludes 105, indicating that the effect may be important (as well as being statistically significant).

Such a procedure avoids the problem of multiple significance testing whenever new data are collected, and avoids the associated corrections for Type I errors (e.g. Bonferroni, discussed in the next chapter and which would make the procedure much less powerful).

5.9   Likelihood versus null hypothesis testing

One might hope that, for a given set of data in a given scenario, different approaches to statistical inference would lead to similar conclusions. In general this is true. However, there are some inconsistencies between different approaches which indicate that all is not well in statistical inference. The likelihood approach was mentioned at the end of Chapter 3, and we revisit this approach here at the end of this chapter in the context of decision making and statistical power. In Neyman–Pearson hypothesis and significance testing approaches, it is claimed that the p value obtained in a statistical test provides evidence against the null hypothesis. The smaller the value is, the less credible is the null hypothesis, since it indicates the probability of obtaining data as extreme (or more extreme) were the null hypothesis true. However, since we are always at the mercy of the possibility of Type I errors, we need to guard against them happening in our study.

Whether we use a one-tailed or two-tailed significance test will affect the p value we would report. A one-tailed test value would be half that of a two-tailed test. A one-tailed test would also be more powerful (a smaller Type II error probability). Whether we just do a single statistical test on a set of data and report the result, or we decide to do more than one test (e.g. if we collect more data to add to existing data after statistical testing, or we perform multiple tests on one set of data), will also make a difference as to whether we would reject or not reject the null hypothesis. Both of these issues (one-tailed versus two-tailed testing, multiple testing) are decided by the investigator. This means that our p value is not the completely objective number we assumed. It depends on the intentions and manoeuvrings of the investigator, in other words, it is to some extent subjective. Royall (1997) writes that p values "are not determined just by what the experimenter did and what he observed. They

depend also on what he would have done had the observations been different – not on what he says he would have done, or even on what he thinks he
would have done, but on what he really would have done (which is, of course,
unknowable)" This means that the evidence provided by the *p* value depends
to some extent upon the provenance of that evidence, and therefore cannot
in all circumstances be used as an objective measure of the strength of evidence. The likelihood approach differs from hypothesis testing (significance
and Neyman–Pearson) and the Bayesian approaches in that it only depends
upon the data collected, nothing more.

We can illustrate how the likelihood approach can conflict with the hypothesis testing approach by considering the sampling distributions in Figures
5.9 and 5.10. In Figure 5.9 the sampling distributions around the null and

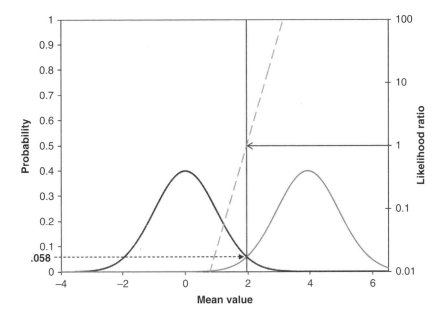

Figure 5.9 Sampling distributions for the mean for the null hypothesis H_0 (left, black) and the
alternative hypothesis H_1 (right, grey). The vertical line represents the criterion, and is positioned
at the beginning of the 5% two-tailed rejection region (an area of 2.5% = α / 2). The overall
Type I error is 5%, as typically used in null hypothesis testing. Because the H_0 and H_1 means
are the same distance from the criterion line, it means that the probability of a Type II error is
2.5%. The likelihoods are where each of the H_0 and H_1 curves hit the criterion line. Here, they do
it in exactly the same place. The horizontal dashed line with arrow indicates this position and
the probability can be read off from the left hand vertical axis, so both have a value of .058.
The likelihood ratio is the ratio of these two values, which of course is 1. The likelihood ratio is
plotted for all the data as the angled grey dashed line. The point at which this hits the criterion
line is indicated by the horizontal line with arrow, and reading off the right hand logarithmic
scale on the right side gives us 1. According to the null hypothesis testing approach, if we had
a mean value at the criterion then we would reject H_0. However the likelihood approach finds no
preference for either hypothesis since the likelihood ratio is 1. Adapted from Dienes (2008).

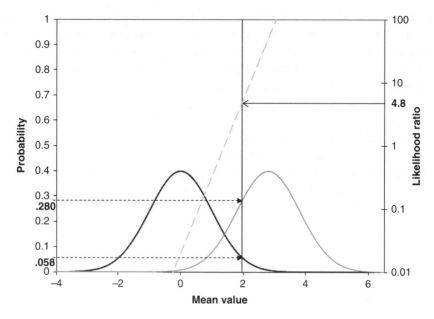

Figure 5.10   The sampling distribution for the mean of H_1 has been moved closer to that for such that the probability for a Type II error β is now 20%. The two horizontal dashed lines with arrows point to where the curves for the two hypotheses hit the criterion value (vertical line). Here, you can see that the probability value for H_1 is higher (.280) compared with the value for H_0 (.058). The ratios of these give us a likelihood ratio of 4.8 in favour of H_1 (horizontal line with arrow). If a sample mean had a value just below the criterion value, then H_0 would not be rejected, while the likelihood ratio would indicate evidence in favour of H_1 (odds of 4.8:1). The actual values given by the probability axis are not important since our statistic is the ratio of these values (giving us the likelihood ratio).

alternative hypotheses are plotted. The distance between the two distributions has been arranged so that α and β are both equal to 5%. The vertical line we can call the criterion line. If a mean is above this value then we reject H_0, else we retain it. The likelihood ratio is calculated by dividing the probability value (left vertical axis) of where the H_1 curve hits the criterion line by the probability value of where the H_0 curve hits the criterion line. You can see that this occurs for the same probability values (both .058), and is indicated by the horizontal dashed line with arrow. Therefore the likelihood ratio is 1 (.058 / .058). The likelihood ratio for both curves is plotted as the angled grey dashed line, and its logarithmic scale is given by the vertical axis on the right. The particular likelihood ratio value for the criterion can be read off on that scale as 1, and represents the point at which the likelihood ratio line (angled dashed) hits the criterion line (horizontal line with arrow). In hypothesis testing a mean obtained at the criterion value would lead us to reject H_0, and claim statistical significance. Yet, according to the likelihood approach the likelihood ratio is 1: meaning that there is equal preference for H_0 and

H_1. It seems peculiar and contradictory that one approach claims an effect is present while the other says there is no preference between the competing hypotheses.

In Figure 5.10, the H_1 mean is shifted closer to the H_0 mean such that β is now 20%. This probability for our Type II error is routinely used in research. The criterion line remains in the same place, and hence α remains at 5%. We now see that the H_0 and H_1 curves hit the criterion line in different places, with readings on the probability scale of .058 and .280, respectively. Making a ratio of them .280 / .058 we get a likelihood ratio of 4.8 (horizontal line with arrow on likelihood ratio scale) in favour of H_1. This is even worse. Just to the left of the criterion line we would not reject H_0 using hypothesis testing, claiming no significant effect. However at this point, according to the likelihood approach we are favouring H_1 by nearly 5:1. Cohen (1977) has suggested that the average statistical power in published studies in psychology journals was 50%. If we adjust our H_1 curve to give a β of 50%, then the likelihood ratio is 6.83. Again, a mean obtained just to the left of the criterion would inform us to claim no significant effect, while the likelihood ratio would almost provide 'fairly strong' evidence in favour of H_1.

The null hypothesis testing approach is to make decisions according to p values obtained in an analysis. If this is accompanied by 95% CIs, then it is often possible to determine whether sufficient data were used (Figure 5.7). However, the likelihood approach also seems to be useful in that it gives us directly the relative evidence for one hypothesis compared with another. It uses information only from the data collected and, through the likelihood ratio, points to the most likely hypothesis. Unless the true value lies exactly between the H_0 and H_1 values (e.g. in Figure 5.9), as the sample size increases, the ratio will be driven ever more strongly in the direction of the hypothesis closest to the truth, whether this be H_0 or a particular H_1 value. In null hypothesis testing, if H_0 is true then the probability of committing a Type I error always remains at 5%.

5.10 Summary

- Statistical power represents the probability of detecting an effect (or the probability of not making a Type II error).

- Power can be calculated *post hoc* to determine whether there was much chance of a study achieving a statistically significant result.

- More often it is used to plan a study, by calculating the sample size, typically assuming power of 80%, $\alpha = .05$.

- The relative costs of Type I and Type II errors should be considered. These will depend on the nature of the study.

- The central limit theorem emphasises the importance of the normal distribution, especially since we are usually interested in the statistical behaviour of **sample** statistics (rather than individual observations).

- There are a number of problems in using p values alone.

- Unless sample size and effect size are taken into consideration, p values can be misleading.

- Hence the use of CIs, typically 95% CIs, are useful in determining whether effects are large enough to be of practical importance and whether enough data have been collected.

- Comparison between the likelihood and hypothesis testing approaches revealed contradictions in that the likelihood ratio could point towards H_1 while hypothesis testing would state there was no statistically significant effect.

- The likelihood approach is consistent with the actual data collected, and provides a relatively unambiguous measure of the strength of evidence for one of two competing hypotheses. Finally, it is unaffected by subjective decisions made by the investigator.

References

Armitage, P., Berry, G. and Matthews, J.N.S. (2001) *Statistical Methods in Medical Research*, 4th edition. Wiley-Blackwell. ISBN: 978-0632052578.

Cohen, J. (1977) *Statistical Power Analysis for the Behavioral Sciences*. Academic Press.

Dienes, Z. (2008) *Understanding Psychology as a Science: An Introduction to Scientific and Statistical Inference*. Palgrave Macmillan.

Reichardt, C.S. and Gollob, H.F. (1997) *What if there were no significance tests?* In: Harlow, L.L., Mulaik, S.A., Steiger, J.H., editors. Mahwah, NJ: Lawrence Erlbaum Associates, p. 273.

Rosenthal, R., Rosnow, R.L. and Rubin, D.B. (2000) *Contrasts and Effect Sizes in Behavioral Research*. Cambridge University Press, p. 4.

Rouder, J.N., Speckman, P.L., Sun, D., Morey, R.D. and Iverson, G. (2009) Bayesian t tests for accepting and rejecting the null hypothesis. *Psychonomic Bulletin & Review*, 16(2), 225–237. Available at: http://pcl.missouri.edu/bayesfactor (accessed 31 August 2013).

Royall, R.M. (1997) *Statistical Evidence: A Likelihood Paradigm*. Chapman & Hall/CRC Monographs.

6

Comparing Groups using *t*-Tests and ANOVA – 'To Compare is not to Prove'

6.1 Aims

The quote is an old French proverb. This chapter will deal with comparing sample means. A common mistake that students make when learning statistics is to state that a test outcome 'proves' that there is a difference between groups. We aim to dispel this myth and to emphasise the concepts of the null and alternative hypotheses. We will deal with *t*-tests and ANOVAs in this chapter. For *t*-tests we will emphasise their usefulness for comparison of two groups and introduce the concept of multiplicity of *p* as a natural progression to the use of ANOVA to test multiple means. We will describe the basic concept of ANOVA – the partitioning of 'variability' (variance) into that which is attributable to the treatment and that which is due to all other possible causes and use illustrations (right-angled triangle) to simply illustrate this concept. The differences between one-way and two-way ANOVA will be described.

Thus far in this book we have established that whenever we collect data we will have variability in the response or whatever it is we are measuring. We need to have some idea of the magnitude of this variability, so we can usually summarise our data using descriptive statistics and visualise them using graphs. It is also informative to have some idea of how the data are distributed, do they conform to a normal distribution or not? However, what we often want to know is whether there is an effect of a treatment that we have applied. Note that we use the word 'treatment' here in the broader statistical sense, meaning a factor that we hypothesise effects a change in a measurement and

Starting Out in Statistics: An Introduction for Students of Human Health, Disease, and Psychology
First Edition. Patricia de Winter and Peter M. B. Cahusac.
© 2014 John Wiley & Sons, Ltd. Published 2014 by John Wiley & Sons, Ltd.
Companion Website: www.wiley.com/go/deWinter/startingstatistics

not in the narrower medical or biological sense, meaning the application of a drug or chemical. We would like to know, given that our data are inevitably variable, how probable it is that the observed effects have arisen by chance alone. We have explored the meaning of statistical probability, how it can be calculated and how we can ascribe a value of probability to our results.

6.2 Are men taller than women?

Let us now return to a very simple example that we have used previously: human adult height. The question we will ask is: 'Does the height of males differ from that of females'? First note that we have specified adult, why? Well one source of variability in measurements of height is growth. Children are still growing and boys' growth patterns are not the same as those of girls (girls start their pubertal growth spurt earlier and tend to reach their adult height at a younger age). Adults on the other hand have reached their maximum height. What other factors might affect height? We know that no single gene determines a person's height; it is influenced by the contributions of many genes, hundreds of variations at more than 180 loci are known (Lango Allen *et al.*, 2010). Food is another important factor: poor nutrition in childhood might restrict the height reached as an adult. These are just a few known factors that affect height, but there are likely many more. We call sources of variation that produce differences in our measurements but are not the factor in which we are interested, *random factors*. We can never *know* absolutely all the random factors that might affect our measurements. We have thought about a few that might affect adult height, but any number of events may have occurred during growth that either promoted or restricted the ability of an individual to reach their potential maximum: spinal curvatures such as scoliosis, childhood illnesses, hormonal imbalances, to name but a few. In old age, adults tend to decrease in height, as a result of musculo-skeletal diseases and atrophy, so we may wish to restrict our sampling to include only adults below 50.

So, we have established that many factors, known and unknown, affect the final height reached as an adult, but we are interested in whether there is a difference between the two genders that is detectable in spite of all the other sources of variation that affect height. We observed in Chapter 4 (Figure 4.4) that a histogram of female adult height is both symmetrical and bell-shaped indicating a normal distribution. Now let us take a look at the histogram for male and female students together (Figure 6.1). This cartoon is also based on Joiner's living histograms (1975). Note that this time rather than a normal distribution, which is what we would observe if each gender is plotted separately, the histogram for both genders together exhibits two peaks. Might this be because male and female heights differ, or did we happen to obtain a sample with a bimodal distribution by chance?

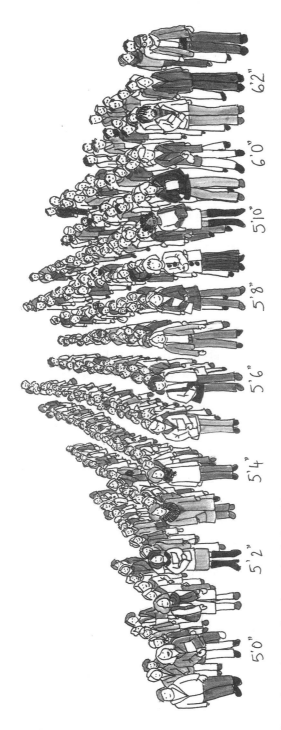

Figure 6.1   Cartoon based on Joiner's (1975) living histograms illustrating the distribution of the heights of a combined sample of male and female college students.

Let us go back to our original question, which was: 'Is there a difference in height between men and women?'. This is a very general question but we have two samples, one for each gender, comprised specifically of students at Pennsylvania State University in 1975. We want to use our samples to make inferences about gender differences in human height in general. This may or may not be a valid proposition. It is possible that our samples reflect the variability in values of height that we would obtain if we sampled randomly both outside and within the university, or it may be that university students are a peculiar breed who differ from other mortals (in their bizarre eating habits, for example) and our samples are not representative. So considering what we have learned from previous chapters, the first thing we must think about is whether our measurements are representative or have we introduced bias by our sampling strategy. In Chapter 1, we learned that we very rarely obtain measurements for the whole statistical population, so we have to use samples as surrogates for populations. We want to infer the characteristics of the population from the samples.

Population parameters are conventionally represented by Greek letters:

μ (pronounced mew) is the symbol for the *population mean*.

σ^2 (pronounced sigma squared) is the symbol for the *population variance* and σ for the *population standard deviation*.

Sample statistics, which we introduced in Chapter 4, are conventionally represented by Roman letters:

$\bar{x}$ (pronounced x-bar) is the symbol for the *sample arithmetic mean*.

s^2 is the symbol for the *sample variance* and s is the symbol for the *sample standard deviation*.

We are now going to assume that our samples of university students' heights are representative of human height in general and put our question in more statistical terms: 'Are the heights of adult women and adult men sampled from two *different* populations of height, one for each gender, for which the population means differ?' Let us emphasise that our data are *samples* of heights, we know neither the value of the population mean(s), nor the population standard deviation(s). So how do we get from samples to populations? We discussed in the previous chapter that the standardised normal distribution has some mathematically interesting properties, one of which is that 95% of the observations occur between 1.96 standard errors above and below the mean. We explored the *z*-test and decided that if our sample mean falls in one of the tails that makes up the remaining 5% (2.5% on each side of the distribution), we can conclude that it is sufficiently distant from the estimated population mean to belong to a different population altogether. The z-test works just fine when the population variance is known or the sample size is large (in excess of 60 is a common recommendation), but with these limitations it isn't always suitable so we need an alternative test. Before we go any further then, let us revisit the central limit theorem, which we discussed in the previous chapter. We are going to do a thought experiment; we'll suppose that the population mean for adult male height is 70 inches (5 feet and 10 inches) with a standard error of the population mean of 3 inches. Now, we will select 50 men at random and measure their heights. The $\bar{x}$ for this sample is 70.1 inches and s is 2.9 inches. We'll repeat the process with a different random sample of 50 men and again record $\bar{x}$ and s. We continue until we have 30 samples for 50 men.

6.3   The central limit theorem revisited

Now let's turn the thought experiment into an empirical one; using Minitab we have sampled randomly from a normally distributed population for which we have conveniently arranged the arithmetic mean to be 70 inches (178 cm) and the standard deviation to be 3 inches (you can try this for yourself, many statistical software packages are able to generate a population of known parameters). Next, the software calculated the descriptive statistics for each (Table 6.1). Now the interesting thing to note is that an awful lot of sample means are pretty close to 70, with standard deviations close to 3. Twenty

Table 6.1 Arithmetic mean and standard deviation of 30 samples of human male height randomly sampled from a population with a mean of 70 inches (178 cm) and standard deviation of 3 inches (8 cm), $n = 50$

Sample number	$\bar{x}$	s	Sample number	$\bar{x}$	s
1	70.2	3.5	16	70.0	2.5
2	70.5	2.7	17	69.6	2.7
3	69.8	2.8	18	70.2	3.0
4	69.5	3.4	19	69.4	3.2
5	70.2	2.8	20	70.3	2.8
6	70.5	3.1	21	70.0	3.6
7	70.3	2.4	22	70.7	3.3
8	70.7	3.0	23	69.4	3.0
9	70.1	2.6	24	70.4	3.1
10	69.9	2.8	25	70.5	3.0
11	69.5	2.8	26	69.6	3.0
12	70.4	2.9	27	70.0	3.3
13	70.1	3.2	28	69.6	3.2
14	69.7	2.7	29	70.2	3.3
15	70.2	3.5	30	70.1	3.1

The data were generated in Minitab.

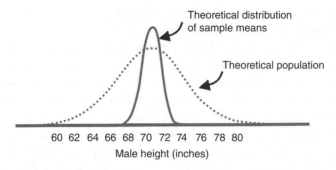

Figure 6.2   A theoretical distribution of a population of male heights with a mean of 70 inches and standard error of 3 inches (dotted line) and a theoretical distribution of the sample means for a large number of repeated random samples of 50 male heights taken from the population (solid line).

six out of the 30 mean values fall within half an inch either side of 70 (i.e. from 69.5 to 70.5). If we repeated this we would get a similar result. What this demonstrates is that we have a large enough sample size, our sample mean and standard deviation are a pretty good estimate of the population parameters in most cases. This might seem quite remarkable, but it does make intuitive sense. If you sample randomly, because most observations (68%, or just over two thirds) occur within one standard deviation above and below the population mean, then repeated samples are more likely to contain those observations than extreme values. Think of it this way: Robert Wadlow, who died in 1940, was the tallest ever recorded male human at 8 feet and 11 inches while the shortest, Chandra Bahadur Dang, is 21.5 inches tall (source: Guinness World Records). Now think about how many men you meet who are close to these two heights. It would be a very rare occurrence; most men will be neither extremely tall nor extremely short. Even a male who is shorter than 5 feet would be fairly unusual.

   If we were to take many more samples from the population than the 30 we have here and plotted their means and compared the distribution with that of our population, we would get something that looked rather like Figure 6.2. If we had an infinite number of sample means, the mean of these would be very close to the population mean but its distribution would be narrower than that for the population. This also makes sense; we would need an awful lot of very small men in a sample size of 50 to obtain a sample mean of 60 inches (152 cm), so there will always be less variation in sample means than there is in the population.

6.4   Student's *t*-test

The simulation above is a demonstration of the properties of the central limit theorem that *in the majority of cases* a sample mean will be within one standard error of the population mean. So that is how we get from a sample to a population. That's all very well, but returning to our original question, we want to know if there is a gender difference in height, how do we do that? In the same way that the descriptive statistics, $\bar{x}$ and s^2, of the male height sample are estimates of the male population parameters μ and σ^2, the female sample descriptive statistics are estimates of the female population parameters. So if we compare the sample means, taking the dispersion into account, we could work out if there is a big enough difference between them to declare either that the two population means differ or alternatively that they are estimates of the same population mean.

When students start to learn statistics, they often find it very confusing that we test a null hypothesis. Why can't we simply test the hypothesis that there is a difference in height between adult men and women? Why do we have to express it as a null? Using a null hypothesis is a rather clever way of arranging the statistical test to compare two sample means even though we do not *know* the value of the means for the populations from which these samples have been drawn. Think of it this way, if there is actually no difference between adult male and female heights, then subtracting the population mean for males from that of females will always give the answer zero. This way we don't ever need to know the absolute value of a population mean; it could be 67 inches, for example, but if it's the *same* population for males and females $67 = 67$ and $67 - 67 = 0$. In mathematical shorthand we can express this $\mu_1 = \mu_2$ and $\mu_1 - \mu_2 = 0$. So our null hypothesis is that the difference will be zero that is, there is no difference in the mean height of adult men and women. Now if there really is a difference, then subtracting the means will produce a value other than zero, and the bigger the difference the further away the value will be from zero.

The principle is exactly the same as we encountered for the z-test, for not only do sample means drawn repeatedly from *one* population exhibit a normal distribution, which is demonstrated by the central limit theorem, so do the *differences in mean values of two samples*. To explain this let's do another thought experiment. This time instead of just taking samples of male heights and calculating the means (Table 6.1) and plotting the distribution of means, we are now going to randomly sample 50 male heights and also 50 female heights. We'll subtract the mean for the females from the mean for the males. This is the difference in sample means. If we continue to repeatedly sample male and female heights and each time calculate the differences in the sample means, these *mean differences* will also conform to a normal distribution, at least it will when we have a large number of pairs of samples. This principle can be empirically simulated using statistical software as we did for the male means in Table 6.1, but it is a bit tedious as we would need a very large number of samples of the differences in means, probably in excess of 100. So you could try this for yourself, but for our purposes here, you will have to take our word for it that this is the case.

Let's just quickly review the story so far:

1. We have discovered that a sample mean, $\bar{x}$, is an estimate of the population mean, μ, and its standard deviation, s, is an estimate of the population standard deviation, σ.

2. If we have two samples, such as male and female heights, their respective descriptive statistics are estimates of the parameters of their respective populations.

3. We want to know if males and females differ in height so to do this we wish to test whether our two samples have the same population mean or whether their population means differ.

4. We conveniently arrange the population mean to be zero on the basis that if the mean population mean for height is the same as the female one, there is zero difference between them.

5. We know that for an infinite number of samples the distribution of the differences in sample means is normal and the mean of these sample means will equal the population mean, zero.

So we could use the same principle as we did for a *z*-test in Chapter 5; as the differences in sample means also exhibit the properties of a normal distribution we could try to work out how many standard errors away from zero difference are the difference in the means of our two samples of height. But to do a *z*-test we need the population variance and we don't have it. So although we have rather cleverly worked out a population mean we have no measure of dispersion; at the moment we have no idea how narrow or broad our distribution of sample means is and unless we know that, we can't work out what are the values the define the cut-off for the tails, 2.5% in each (Figure 6.3).

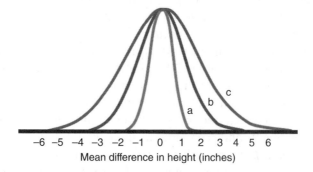

Mean difference in height (inches)

Figure 6.3 Some possible distributions for the differences in sample means, that is, male height minus female height. Although we know the population mean is zero, as is the mean of the possible distributions a, b and c, unless we have a measure of dispersion we cannot tell whether the distribution is very narrow as in a, wider as in b, or wider still as in c. Unless we know the standard error of the distribution we cannot work out how far away from zero the tails lie.

Just as a measure of central location is meaningless without an accompanying measure of dispersion (see Chapter 4), calculating the difference in two means (by subtracting one from the other) is equally unhelpful unless we factor in the dispersion of the differences between the two samples.

In a flash of inspiration we might ask 'But we have the sample variances could we not use those to estimate the standard error of the differences'? This seems like a sensible solution but there is a slight problem with this as first noted by William Sealy Gosset (working for the Guinness brewery at the time): using the sample standard error of the difference to describe the dispersion in the distribution of the differences in means did not result in a normal distribution. The 'tails' were extended compared with the normal distribution and crucially the shape of this distribution changed with sample size (Figure 6.4). The smaller the sample size, the bigger and more extended were the tails. At very large sample sizes the shape was indistinguishable from a normal distribution. This Student's *t*-distribution, as it became known, was named after the pseudonym William Sealy Gosset used for publication as his employment at the Guinness brewery prevented him publishing under his own name.

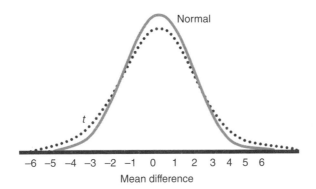

Figure 6.4   Diagrammatic representation of the *t* distribution for a small sample size relative to a normal distribution. The tails of the *t* distribution are more extended and the distribution is somewhat flatter.

So for small sample sizes, or even for larger ones where we use the *sample variances* as the estimate of dispersion, if we were to use a normal distribution to describe the distribution of the difference in means for the statistical test, falsely rejecting the null hypothesis (Type I error) would occur more frequently than five times in 100 tests. So if the sample variance is used, as we will have to do in this case, then the differences in means have to be greater to attain the same level of significance – this is what in effect the *t*-distribution does. Thus, a *t*-test compensates for using smaller sample sizes by making the test rather more stringent than a *z*-test.

For Student's *t*-test, we subtract one sample mean from the other to use as an estimate of the difference in population means, and then we divide this difference by an estimate of dispersion in the population, called the *standard error of the difference* (Equations 6.1 and 6.2). This is the point where you probably inwardly scream 'Oh no, not equations', but although they might look horrendous, they are actually not as complicated as they look at first glance.

Where we have good reason to assume that the population variances are equal (see later section titled assumptions of the *t*-test) we can estimate a common or pooled standard deviation, $s_{x_1} s_{x_2}$, which is what we will use here (Equation 6.1). This is calculated from the weighted average of the two sample variances: that is, rather than simply add the two variances together and take their square root to obtain a standard deviation, we also take into account the size of each sample (n_1 and n_2).

Equation 6.1   Formula for the pooled standard deviation. The subscripts 1 and 2 refer to the first and second samples respectively, n is the number of observations in a sample, s^2 is the sample variance – we learned how to calculate this in Chapter 4.

Take square root for
pooled standard
deviation

$$s_{x_1 x_2} = \sqrt{\frac{(n_1 - 1)s_1^2 + (n_2 - 1)s_2^2}{n_1 + n_2 - 2}}$$

Weighted average of
the two sample
variances

If you would like to calculate the sample variances yourself, follow Steps 1–4 for the standard deviation example in Chapter 4; but they are 8.246 for the males and 6.398 for females. There are 109 male heights and 123 female heights, so 231 observations in total. We can use these numbers to work out the pooled standard deviation.

6.4.1   Calculation of the pooled standard deviation

Step 1.   Calculate the value of the numerator (top part of the fraction):

$$(109 - 1) * 8.246 + (123 - 1) * 6.398$$
$$= 108 * 8.246 + 122 * 6.398$$
$$= 890.568 + 780.556$$
$$= 1671.124$$

Step 2.   Calculate the value of the denominator (bottom part of the fraction):

$$109 + 123 - 2 = 229$$

Step 3. Divide the numerator by the denominator:

$$1671.124 / 229 = 7.2973$$

Step 4. Take the square root of the value obtained in Step 3:

$$s_{x_1} s_{x_2} = \sqrt{7.2973} = 2.713$$

To obtain *t* first we subtract the arithmetic mean of the second sample from that of the first sample, $\bar{x}_1 - \bar{x}_2$, this is the difference in means. Next we obtain the standard error of the difference and divide it into the difference in means. The standard error of the difference is calculated by multiplying the pooled standard deviation by the square root of the sum of the sample size reciprocals. If you are lost, don't worry as we'll go through the calculations step by step below. If we have a large value for the standard error of the difference relative to the difference in means then we have a lot of variability in the data and *t* will be smaller. We call *t* the test statistic. Conversely if we have a small value for the standard error of the difference relative to the difference in means then we have a lower variability in the data and *t* will be larger.

Equation 6.2 Formula for Student's *t*-test with unequal sample sizes. The subscripts 1 and 2 refer to the first and second samples respectively, $\bar{x}$ = the sample arithmetic mean, n is the number of observations, $s_{x_1 x_2}$ is the pooled sample standard deviation (see Equation 6.1).

$$t = \frac{\bar{x}_1 - \bar{x}_2}{s_{x_1 x_2} * \sqrt{1/n_1 + 1/n_2}}$$

Subtract the sample means from each other – this estimates the difference in population means

Divide by the standard error of the difference – this accounts for dispersion (variability in the data)

6.4.2 Calculation of the *t* statistic

Step 1. Calculate the difference in means (it doesn't matter which we call sample 1 or 2, but here we'll call the sample for males sample 1):

$$70.046 - 64.713 = 5.333 \text{ inches}$$

Step 2. Sum the two sample size reciprocals (1 / *n* in Table 6.2):

$$0.00917 + 0.00820 = 0.01737$$

Table 6.2 Statistics required to calculate the value of *t*

	Men		Women
Arithmetic mean	70.046		64.713
n	109		122
1 / *n*	0.00917		0.00820
$s_{x_1 x_2}$		2.713	

The units for the mean are inches.

Step 3. Take the square root of the value obtained in the previous step:

$$\sqrt{0.01737} = 0.1318$$

Step 4. Multiply the value obtained in Step 3 by the pooled standard deviation:

$$0.1318 * 2.713 = 0.3576$$

This is the standard error for the difference in means

Step 5. Divide the difference in means obtained in Step 1 by the standard error for the difference in means from Step 4 to obtain *t*:

$$t = 5.333 / 0.3576 = 14.91$$

6.4.3 Tables and tails

Our value of *t* for the difference in means is 14.91 but what does this signify – is there a difference or not? The statistical significance for a *t*-test requires one step further than a *z*-test – we have to account for the sample size. Once upon a time we would need to manually look up our value of *t* in a *t*-table, and if our value of *t* turned out to be greater than the value given in the table, which was called the *critical value,* for the significance level (*α*), usually .05, we would say that the result was a statistically significant difference between the two means. These days the significance, expressed as an exact probability value, is calculated automatically by statistical software.

Before we reveal the results of our *t*-test, let's take a look at part of a *t*-table (Table 6.3). For Student's *t*-test the degrees of freedom (df) (i.e. the number of independent pieces of information used to estimate the population parameters) is *n* − 2 because we subtract one observation from each *group* and we have two groups. Note that we have two rows of *α*, one for a

Table 6.3 Statistical table for the *t* distribution for 2–10 degrees of freedom

	α (1 tail)	.050	.025	.010	.005	.001	.0005
	α (2 tail)	.100	.050	.020	.010	.002	.001
Degrees of freedom	2	2.920	4.303	6.965	9.925	22.328	31.598
	3	2.353	3.182	4.541	5.841	10.215	12.924
	4	2.132	2.776	3.747	4.604	7.173	8.610
	5	2.015	2.571	3.365	4.032	5.893	6.869
	6	1.943	2.447	3.143	3.707	5.208	5.959
	7	1.895	2.365	2.998	3.499	4.785	5.408
	8	1.860	2.306	2.896	3.355	4.501	5.041
	9	1.833	2.262	2.821	3.250	4.297	4.781
	10	1.812	2.228	2.764	3.169	4.144	4.587
	∞	1.645	1.960	2.326	2.576	3.090	3.291

The shaded value is the critical value for the *t*-test performed for male versus female student height.

one-tailed test and the other for a two-tailed test. We should use a two-tailed test if differences can be either negative (smaller than a mean of zero) or positive (larger than a mean of zero). This is the case for most biological data. In our example of height differences, women could be either taller or shorter than men, so we use the two-tailed test. Very occasionally, observations can change in only one direction, either increase or decrease or we may be interested in only one direction of change. In general this is rare in biological data so a two-tailed test is that which you will use most often.

To return to our *t*-table, if we were testing for a difference in means for two groups with five observations in one and six observations in the other, our degrees of freedom would be $5 + 6 - 2 = 9$. We decide upon using a 5% probability level and a two-tailed test, so we look up the value of *t* for 9 degrees of freedom in the column .05 (two-tailed). Critical *t* is 2.262, so our calculated value of *t* has to be bigger than this for us to be able to say that the difference in means is large enough that its value falls in one of the tails of the distribution. If this were the case, we would report the probability as $p < .05$. However, as we can now obtain exact probabilities we should report them as such and avoid using > or <. If our variable could change in only one direction, then we would select a one-tailed test and we would use the column headed .05 in the top row of Table 6.3.

The output from the *t*-test using Minitab is reported in Box 6.1. First we have the sample descriptive statistics for the two samples: n, $\bar{x}$ and s. The standard error of the mean is the statistic that can be used to calculate the 95% confidence intervals for the population mean of each gender and tells us the range of values in which the population mean is likely to be located in 95 tests out of 100 (see Chapter 5).

Box 6.1   Minitab output for two sample t-test for men versus women

	Men	Women
n	109	122
Mean	70.05	64.71
Standard deviation	2.88	2.55
Standard error of the mean	0.28	0.23

```
Difference = μ male - μ female
Estimate for difference: 5.333
95% confidence interval for difference (4.628, 6.037)
t-test of difference = 0 (vs not =):
t-value 14.91; p-value = 0.000; DF = 229
Both use pooled standard deviation = 2.71
```

The difference assumed to be zero under our null hypothesis is the male population mean minus the female population mean, and the *estimate* of the actual difference is the male sample mean minus the female sample mean, which is 5.33 inches, so the test results declare that on average, women are just over 5 inches shorter than men. The 95% confidence interval tells us the location of the population mean for the difference lies within 4.628 to 6.037 inches at a probability of 0.95. That's quite a long way from zero difference and even four and a half inches is a not a small height difference. The effect size (difference in means divided by the pooled standard deviation see Chapter 5) is 5.333 / 2.71 = 2.05, which is quite large. On the following line we have the test statistic, *t*, and value of *p* for the null hypothesis that the difference in population means is zero (versus the alternative hypothesis that it is not equal to zero). Probabilities lower than 0.001 (less than one chance in a thousand) are usually reported as 0.000 by statistical software, that is to only three decimal places. In publications it should be written as < .001. The degrees of freedom for our *t*-test is the number of observations in total minus one for each group ($n_1 + n_2 - 2$) or $109 + 122 - 2 = 229$.

Interestingly, our samples collected in 1975 are very close to the mean heights of men and women of similar age in the United Kingdom and the United States in the first decade of the twenty-first century. In the United Kingdom men measured 69.6 inches and women 64.5 inches, and in the United States men measured 69.9 inches and women 64.3 inches, $n \approx 500 - 800$ (McDowell *et al.*, 2008; HSCIC, 2011).

---

t-test results summary: Mean height of male students at Pennsylvania State University was greater by 5.33 inches than that of female students (two-tailed independent *t*-test, $t_{229} = 14.91$, $p < .001$) with a 95% confidence interval for the difference (4.628, 6.037).

This type of *t*-test is also called a two-sample *t*-test or an independent *t*-test. Independent in this context means that the two samples of heights are from different individuals and that the observations in one group are not influenced by (i.e. are independent of) the observations in the other. When the sample sizes are equal, the formula for the calculation of the pooled variance is slightly different, but this is a technical detail and does not change the fundamental principles of the test that we have described above.

6.5 Assumptions of the *t*-test

The Student's *t*-test is a fairly robust and pretty powerful statistical test, but it does require that certain assumptions are met for the test to be valid. These are:

1. The samples are drawn at random from the population.

2. The population distributions are approximately normal.

3. The population variances are equal.

This first assumption is the trickiest to determine because there is no test that can help us to decide whether or not it has been met. Ensuring that the samples are drawn at random from the population has much to do with the experimental design. We also need think very carefully about the null hypothesis that we are testing. For sampling to be random, each object to be sampled must have an equal chance of being selected as any other. Imagine that we wanted to compare some characteristics of Neanderthals with modern humans. Neanderthal skeletons are not abundant and found mostly in museums. And that is the key point – museum specimens may have ended up there for particular reasons; they were found in accessible locations and were better preserved because of the environment in which they were found, such as a dry cave or a peat bog. So museum samples may be unrepresentative of Neanderthal skeletons.

The assumption of normality may be checked by plotting a histogram (frequency distribution) and comparing its shape with that of a bell-shaped curve, but this method will not work well with small sample sizes. Another graphical method is to use a quantile–quantile plot (Q–Q plot), available in many statistical software packages. A quantile is a division of a frequency distribution into equal ordered subgroups. To construct a Q–Q plot, the observations are converted to quantiles and a *z*-score corresponding to each quantile is plotted on the *x*-axis with *z*-scores for the expected quantiles of normally distributed data on the *y*-axis (Figure 6.5). If the data were perfectly normally distributed, then they would fall exactly on the line $x = y$ (solid grey line in the figure). The further the data are from a normal distribution, the greater

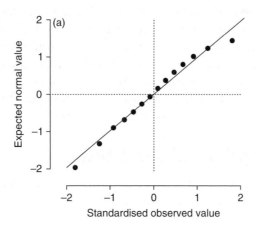

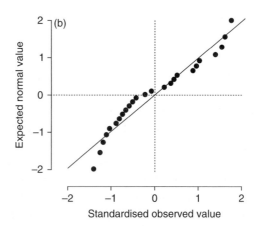

Figure 6.5   Quantile–quantile plots (Q–Q plots). The solid line is $x = y$. (a) Height of male students at Pennsylvania State University in 1975 which are normally distributed and (b) Heart rate in undergraduate students from the University of London in 2012. Q–Q plots are readily constructed by statistical software packages such as SPSS and recent versions of Minitab. Briefly, the software first ranks the observations and the quantile is the midpoint of each ranked sub-group, for example, the first midpoint of the first quantile is equal to $(1 - 0.5) / n$. For example, for the ranks of 10 observations, the first midpoint would be $(1 - 0.5) / 10 = 0.05$, the second midpoint would be $(2 - 0.5) / 10 = 0.015$, and so on. For the theoretical values of the normal *z*-score, it finds the area of the distribution in the left tail corresponding to the value of the quantile midpoint and plots each of those against the computed *z*-score for the observation.

the deviation of the points on the Q–Q plot from the line $x = y$. The observations for male height are normally distributed as the observed versus expected quantiles deviate little from $x = y$, but the observations for systolic blood pressure are not normally distributed as many of the observed versus expected quantiles do not fall on or near the line $x = y$. Lastly, a number of statistical tests are available to test whether data conform to a normal distribution or not. These all work in slightly different ways, and it is beyond the scope of this book to discuss the relative merits of each, but in essence these tests return a probability value which can be used to decide whether or not to reject the null hypothesis that the distribution of the sample data does not differ from a normal distribution. If the probability value returned by the test is large ($\geq .05$) then we fail to reject the null hypothesis and hence the sample distribution is consistent with a normal distribution. If $p < .05$, however, we would reject the null hypothesis and the sample distribution does differ from a normal distribution. In this case, the data do not meet the assumptions of the test and we would need to either transform the data (discussed at the end of this chapter) or use a different statistical test that does not rely on a normal distribution. A comparison of four commonly used tests of normality concluded that the Shapiro–Wilk test was the most powerful, followed by Anderson–Darling and Lilliefors tests (Razali and Wah, 2011). The least powerful of the four was the Kolmogorov–Smirnov test, but with small sample sizes (<30) the power of all four tests was low. Note, however, that with large sample sizes (>100) the tests become too powerful and detect small, unimportant deviations from the normal distribution.

Similarly, the third assumption of equal variances can be tested using a statistical test called Levene's test. Again, this test is widely available in statistical packages and compared favourably against six other tests of equality of variances in a study by Lim and Loh (1996). Levene's test for equality of variances compares the dispersion of the data for two samples and returns a probability value, which is used to reject or fail to reject the null hypothesis. The null hypothesis is that the variances are equal, so a probability of $\geq .05$ signifies that they are equal whereas $p < .05$ is indicative that they are not equal. Fortunately there is a variation of Student's *t*-test, called Welch's *t*-test that can be used when the variances are not equal. In Welch's *t*-test the dispersion of each sample is estimated separately rather than pooled and the degrees of freedom is reduced which increases the stringency of the test to compensate for unequal variances.

6.6 Dependent *t*-test

A variation of the *t*-test can also be used when two sets of observations are not independent, but are related to each other in a pairwise fashion. This is called a *dependent* or *paired samples t*-test. For example, if we measure a variable in the **same** individuals before and after an intervention such as a drug, then the two measurements cannot be independent of each other. In this case we are interested in the difference between the second measurement and the first measurement. In effect we have one sample, which is the difference between the two observations. We take the mean of all the differences as the sample mean and test the null hypothesis that it does not differ from zero. If the differences between the two measurements are the same or only slightly different in all the individuals we measured, then the sample mean will be close to zero. If the measurements are very different before and after, say all much lower or all much higher, then the differences will be large and the sample mean will be further from zero, either negative or positive. As the dependent *t*-test effectively uses one sample, the differences, the assumption of equal variances does not apply, but the other two assumptions discussed above still do.

6.7 What type of data can be tested using *t*-tests?

t-Tests are most appropriate for quantitative continuous variables that meet the assumptions of the test. Data that are counts (discrete quantitative variables) generally do not meet the assumptions of normality because they can take only positive values and are usually skewed. Often they follow a Poisson distribution (many low values and few high values) such as the example of eosinophils in blood in Chapter 4.

6.8 Data transformations

Where data are not normally distributed either a non-parametric alternative may be used (see Chapters 9 and 10) or the data may be transformed and tested again for normality. While not restricted to *t*-test data, we briefly discuss transformation here as it is the first time we have encountered the challenge of testing data that are not normally distributed. Common transformations are log, square root and arcsine. Transformation simply involves taking the log, square root, etc. of the raw data and performing the statistical test on the transformed values, as long as they comply with a normal distribution, instead of the original ones. Transformation changes the scale of measurement; it is not 'fiddling' with the data as some people worry it might.

For proportions, or percentages derived from counts expressed as proportions, an arcsine transformation is suitable. Positively skewed data may be transformed by taking the square root (particularly useful for counts which often exhibit positive skewness). Log transformations may be used for continuous variables. For the Excel commands used to perform these transformations, please refer to the basic maths refresher section at the front of this book. Note, however, that some transformations will not work with certain numbers as you cannot take a log or square root of zero or negative values. One option for transforming zero and negative values is to first add a constant that will make them all positive, for example, if biggest negative number is −9 add 10 to all the observations and then transform. One problem that can sometimes occur with transforming data is when one treatment group is normally distributed while the other is not, say it is skewed. Transforming may normalise the skewed data but at the same time it may also make the normally distributed group data non-normal, so you end up in the same situation only the other way round. In this case a non-parametric test is the only other alternative.

6.9   Proof is not the answer

Before we progress any further we would like to introduce the concept that statistical tests do not provide *proof* of anything. It is a common mistake for students to think that results prove a hypothesis. They do not – to compare is not to prove. Milton Friedman, an American economist and statistician, once wrote 'the only relevant test of the validity of a hypothesis is comparison of its predictions with experience'. Hypotheses are predictions. We test their validity by taking samples (experience), comparing them against each other, and finally computing a probability that our results occurred by chance alone. On some occasions, by chance, we will arrive at the incorrect conclusion and for a probability of .05 this is five times out of 100 or one time in 20. The lower the probability value, the less likely we are to make this error, but the *possibility* of making it is always there. Therefore, we cannot *prove* our hypothesis.

6.10   The problem of multiple testing

This brings us nicely on to the topic of multiple testing. When the null hypothesis is true (e.g. there is no difference) the probability of obtaining a significant result (Type I error) is .05, so the probability of not obtaining a significant result is .95. Now if we had three groups, A, B and C, in our experimental design rather than two, would we be able simply to perform a *t*-test for each of the three comparisons: A versus B, A versus C and B versus C? You might be tempted to answer 'Yes, why not?' but now we will demonstrate why this is not a good idea. For three tests the probability of not obtaining a significant result is not .95 for each test but $.95 * .95 * .95 = .86$. So now the probability of Type I error is $1 - .86 = .14$ not .05! The greater the number of comparisons the greater the likelihood of making Type I error and falsely declaring that the null hypothesis is true when it is not.

You may think that you are unlikely ever to need to do multiple tests but in fact the era of genomics and imaging has brought this problem to the foreground. Microarrays, next-generation sequencing data, and various types of brain scanning techniques involve making thousands or tens of thousands of comparisons. The problem of multiple testing is therefore very applicable to modern biological sciences and we will deal with methods for analysing such complex data in a later chapter.

Is there any way around this problem? Well, we could set α at .01 instead of .05 but we would then risk the opposite effect of making a Type II error and falsely fail to reject the null hypothesis (stating that there is no difference when one exists). Alternatively, we could apply something call *Bonferroni's correction*. This reduces the value of α according to the number of comparisons, so for three samples, there would be three comparisons and α would be .05 / 3 = .0167. However, for four samples, there would be 10 *t*-tests and α would be .05 / 10 = .005, so now we would potentially run into the problem of Type II error. All our *p*-values would have to be less than .005 for us to reject the null hypothesis. Bonferroni's correction is conservative, so Type II error is a very real possibility if we need to perform many *t*-tests (and Type I error a possibility if we don't use the correction). If we wish to compare multiple means, the *t*-test is not the best way to go about it. One answer is to use a test that can deal with more than two comparisons without a loss of power – the analysis of variance.

6.11   Comparing multiple means – the principles of analysis of variance

As we have already observed, when we perform an independent *t*-test we test the null hypothesis that $\mu_1 = \mu_2$. When we have multiple means, the null hypothesis is extended to $\mu_1 = \mu_2 = \mu_3 \ldots = \mu_n$, for up to *n* samples, and as few as two. So we are saying that all the means are equal and testing to find out whether we can reject that null hypothesis. The analysis of variance, or ANOVA, will tell us whether two or more means differ from each other. There are many variations of the test but all work on the principle that dispersion in the data can be attributed to two sources: the effect of the factor in which we are interested and random factors.

We will use an example of placebo pills and blood pressure. A placebo is a tablet that contains no active ingredient and so theoretically it should not have any effect. Placebos are widely used as a negative control in clinical trials of new drugs but it is well documented that they often have an effect on the variable being measured when compared with no treatment. The physical characteristics of the placebo play a role in its effectiveness, expensive placebos produce a greater effect than cheap ones, and some colours are more effective than others (Blackwell *et al.*, 1972; Waber *et al.*, 2008). Let us suppose we wish to test the effect of four treatments on systolic blood pressure in mildly hypertensive subjects: no treatment, white, red or blue placebo pills. Note again that the word 'treatments' is used here in its statistical rather than medical sense. 'Treatment' is the fixed factor and we say it has four levels. For simplicity we will use eight age-matched subjects per treatment group and measure systolic blood pressure following 2 weeks

Table 6.4 Systolic blood pressure following administration of either no treatment or red, white or blue placebo pills for 2 weeks to subjects with mild hypertension, $n = 8$ per group

No pills	White pills	Red pills	Blue pills
139	142	151	123
139	133	145	136
144	143	144	125
149	139	150	123
144	123	145	124
143	138	146	130
150	138	156	134
140	134	145	124

of treatment, although if we really were to conduct such a study we would require a much larger sample size (Table 6.4). Because we have used different subjects the four samples are independent of each other.

Systolic blood pressure varies and innumerable transient factors can affect its values including mood, anxiety, time of day, exercise, caffeine, alcohol, and white coat effects. The white coat effect refers to the fact that in some people simply having their blood pressure measured, whether that person is wearing a white coat or not, increases its value. So the variability in blood pressure measurement could be due to any of these things, which we call random factors, or it could be due to the treatment that we apply. We want to know if there is an effect of the factor 'treatment' over and above the variation due to random factors. In the first instance, we simply ignore the fact that we have different treatments and we pool all the observations together. We want to separate the variation due to the fixed factor treatment (meaning all treatments together) from that due to all the random factors, which is everything else. We call the variation in the data due to random factors the *error* or *residual* variation. Here we are going to use the term 'error'.

The theory behind an ANOVA is modelled by Pythagoras's theorem of right-angled triangles, $a^2 + b^2 = c^2$ (Efron, 1978). Before you allow the previous sentence to reduce you to jelly, take a look at Figure 6.6. Now imagine that the area c^2 represents the total variation in the data and its value is 25 squared units, the area a^2 represents the variation due to treatment with a value of 16 squared units and the area b^2 represents the variation due to error (random factors) and its value is 9 squared units, then $16 + 9 = 25$. There is more variation due to treatment (16 squared units) than there is due to error (9 squared units). Where b is short relative to a, its square will be a smaller proportion of the total area (variation) and where b is long relative to a, its square will constitute a large proportion of the total area (variation).

So how do we use a geometric model, Pythagoras' theorem, in analysis of variance? ANOVA assumes that the observations *deviate* from the *grand*

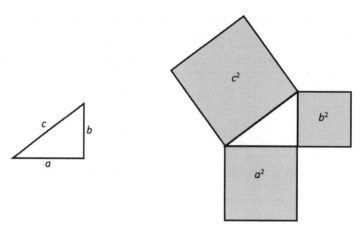

Figure 6.6 Pythagoras' theorem. A right-angled triangle has sides of length a, b and c, with c being the longest side or hypotenuse (left image). If we square each side (multiply the length of each side by itself) we will have three squares, a^2, b^2 and c^2 (right image). If we add together the areas a^2 and b^2 they will equal the area of c^2. We can test this with real numbers. If $a = 4$, $b = 3$ and $c = 5$, then $4^2 + 3^2 = 5^2$ or $16 + 9 = 25$.

mean (overall arithmetic mean) because of variation attributable both to the effect of the fixed factor and random factors. This is how we do it:

Step 1. Calculate the grand mean by pooling all the observations together and ignoring the fact that they belong to different treatment groups:

> The sum of the observations is 4439.
> Grand mean = 4439 / 32 = **138.7188**

Step 2. Subtract the grand mean from each observation – these are the deviations and then square each deviation. Next sum (add up) the squared deviations:

Treatment	Observations – grand mean	Deviations	Deviations squared
No pills	139 − 138.7188 =	0.2812	0.0791
No pills	139 − 138.7188 =	0.2812	0.0791
No pills	144 − 138.7188 =	5.2812	27.8911
No pills	149 − 138.7188 =	10.2812	105.7031
No pills	144 − 138.7188 =	5.2812	27.8911
No pills	143 − 138.7188 =	4.2812	18.3287
No pills	150 − 138.7188 =	11.2812	127.2655
No pills	140 − 138.7188 =	1.2812	1.6415
White pills	142 − 138.7188 =	3.2812	10.7663
White pills	133 − 138.7188 =	−5.7188	32.7047
White pills	143 − 138.7188 =	4.2812	18.3287
White pills	139 − 138.7188 =	0.2812	0.0791
White pills	123 − 138.7188 =	−15.7188	247.0807

White pills	138 – 138.7188 =	–0.7188	0.5167
White pills	138 – 138.7188 =	–0.7188	0.5167
White pills	134 – 138.7188 =	–4.7188	22.2671
Red pills	151 – 138.7188 =	12.2812	150.8279
Red pills	145 – 138.7188 =	6.2812	39.4535
Red pills	144 – 138.7188 =	5.2812	27.8911
Red pills	150 – 138.7188 =	11.2812	127.2655
Red pills	145 – 138.7188 =	6.2812	39.4535
Red pills	146 – 138.7188 =	7.2812	53.0159
Red pills	156 – 138.7188 =	17.2812	298.6399
Red pills	145 – 138.7188 =	6.2812	39.4535
Blue pills	123 – 138.7188 =	–15.7188	247.0807
Blue pills	136 – 138.7188 =	–2.7188	7.3919
Blue pills	125 – 138.7188 =	–13.7188	188.2055
Blue pills	123 – 138.7188 =	–15.7188	247.0807
Blue pills	124 – 138.7188 =	–14.7188	216.6431
Blue pills	130 – 138.7188 =	–8.7188	76.0175
Blue pills	134 – 138.7188 =	–4.7188	22.2671
Blue pills	124 – 138.7188 =	–14.7188	216.6431

Total sum of squares: 2638.4688

We now have the total sum of squares (SS), **2638.4688**. This is the total variation in the data and is represented by area c^2 in Figure 6.6.

Step 3. Calculate the arithmetic mean of each sample (use the data in Table 6.4 if that is easier)

No pills – the sum of the observations is 1148.
Mean = 1148 / 8 = **143.500**

White pills – the sum of the observations is 1090.
Mean = 1090 / 8 = **136.250**

Red pills – the sum of the observations is 1182.
Mean = 1182 / 8 = **147.750**

Blue pills – the sum of the observations is 1019.
Mean = 1019 / 8 = **127.375**

Step 4. For each treatment group, subtract the sample mean calculated in Step 3 from the grand mean we obtained in Step 1, square the answer, and multiply by 8 (as there are eight observations per group)
Using: (grand mean – sample mean)^2 * 8
No pills

$$(138.7188 - 143.500)^2 * 8$$
$$= -4.781^2 * 8$$
$$= 22.860 * 8 = \textbf{182.879}$$

White pills

$$(138.7188 - 136.250)^2 * 8$$
$$= 2.469^2 * 8$$
$$= 6.095 * 8 = \mathbf{48.760}$$

Red pills

$$(138.7188 - 147.750)^2 * 8$$
$$= -9.031^2 * 8$$
$$= 81.563 * 8 = \mathbf{652.501}$$

Blue pills

$$(138.7188 - 127.375)^2 * 8$$
$$= 11.344^2 * 8$$
$$= 128.682 * 8 = \mathbf{1029.454}$$

Step 5.   Sum the four answers obtained in Step 3.

$$182.879 + 48.760 + 652.501 + 1029.454 = 1913.594 \text{ or } \mathbf{1913.6}$$

This is the treatment sum of squares. This is the variation in the data due to the fixed factors (treatment) and is represented by area a^2 in Figure 6.6.

Step 6.   At this point we could cheat and subtract the treatment sum of squares from the total sum of squares to obtain the error sum of squares, but we are going to show you how to calculate the error sum of squares directly from the data. You can check that you get same result either way. For each treatment group, subtract the sample mean from each observation – these are the deviations and then square each deviation – then sum (add up) the squared deviations for that sample:

No pills

Observations – sample mean	Deviations	Deviations squared
139 − 143.50 =	−4.500	20.2500
139 − 143.50 =	−4.500	20.2500
144 − 143.50 =	0.500	0.2500
149 − 143.50 =	5.500	30.2500
144 − 143.50 =	0.500	0.2500
143 − 143.50 =	−0.500	0.2500
150 − 143.50 =	6.500	42.2500
140 − 143.50 =	−3.500	12.2500
		126.0000

White pills

Observations – sample mean	Deviations	Deviations squared
142 − 136.25 =	5.750	33.0625
133 − 136.25 =	−3.250	10.5625
143 − 136.25 =	6.750	45.5625
139 − 136.25 =	2.750	7.5625
123 − 136.25 =	−13.250	175.5625
138 − 136.25 =	1.750	3.0625
138 − 136.25 =	1.750	3.0625
134 − 136.25 =	−2.250	5.0625
		283.5000

Red pills

Observations – sample mean	Deviations	Deviations squared
151 − 147.75 =	3.250	10.5625
145 − 147.75 =	−2.750	7.5625
144 − 147.75 =	−3.750	14.0625
150 − 147.75 =	2.250	5.0625
145 − 147.75 =	−2.750	7.5625
146 − 147.75 =	−1.750	3.0625
156 − 147.75 =	8.250	68.0625
145 − 147.75 =	−2.750	7.5625
		123.5000

Blue pills

Observations – sample mean	Deviations	Deviations squared
123 − 127.38 =	−4.375	19.1406
136 − 127.38 =	8.625	74.3906
125 − 127.38 =	−2.375	5.6406
123 − 127.38 =	−4.375	19.1406
124 − 127.38 =	−3.375	11.3906
130 − 127.38 =	2.625	6.8906
134 − 127.38 =	6.625	43.8906
124 − 127.38 =	−3.375	11.3906
		191.8750

Step 7. Sum the four answers obtained in Step 6:

$$126.0000 + 283.5000 + 123.5000 + 191.8750 = \textbf{724.9}$$

This is the *error sum of squares* (SS). This is the variation in the data due to random factors (error) and is represented by area b^2 in Figure 6.6.

We have now partitioned the total variation into that due to treatment and that due to error.

Step 8. Calculate the variances. In an ANOVA the variances are equivalent to the *mean squares* (MS). The sum of squares must be corrected for the degrees of freedom (df) to produce the mean squares.

The df for the total is the number of replicates − 1, that is, $32 - 1 = \mathbf{31}$
The df for treatments is the number of treatments − 1, i.e. $4 - 1 = \mathbf{3}$
The error df is total df − treatments df, ie. $31 - 3 = \mathbf{28}$

To obtain the mean squares (variances) for the treatments and error we need to divide the sum of squares from Steps 5 and 7 by the corresponding degrees of freedom above:

Treatment MS = Treatment SS/Treatment df = 1913.6 / 3 = **637.9**
Error MS = Error SS/Error df = 724.9 / 28 = **25.9**

The treatment mean squares and error mean squares are different estimates of the population variance. The hypothesis we are testing is whether the variance associated with the treatments is so large, when compared with the error variance that it could not have occurred if the null hypothesis were correct.

Step 9. Calculate the test statistic, *F*:

$$F = \text{Treatment MS/error MS}$$
$$= 637.9 / 25.9$$
$$= \mathbf{24.64}$$

The variance ratio is called *F* and is the test statistic for an ANOVA. We should be able to recognise that when the treatment MS is large and the error MS is small, *F* will be large and most of the variation in the data will be due to treatment so there will be a difference in group means. When the null hypothesis is true the variation from the treatments will be similar in magnitude to that from error and the *F* ratio will be approximately equal to 1. When the treatment MS is smaller and the error MS is larger, *F* will be small and most of the variation in the data will be as much due to random factors as the treatment difference. As the mean squares are the measure of variance then if the null hypothesis is true we would expect the size of variance due to the treatments to be no different from the remainder attributable to random error. In other words, the *F* ratio would be 1 if the null hypothesis were correct.

So, where there is an effect of the treatments we would anticipate that treatment MS would be larger than error MS. But how much larger does it have to be to reject the null hypothesis? As we saw for the *t*-test the critical values of *F* have been tabulated (although nowadays these reside within the computer

Table 6.5 Statistical table for the F distribution for selected degrees of freedom

Numerator df	1	2	3	4	5
21	4.32	3.47	3.07	2.84	2.68
22	4.30	3.44	3.05	2.82	2.66
23	4.28	3.42	3.03	2.80	2.64
24	4.26	3.40	3.01	2.78	2.62
25	4.24	3.38	2.99	2.76	2.60
26	4.22	3.37	2.98	2.74	2.59
27	4.21	3.35	2.96	2.73	2.57
28	4.20	3.34	2.95	2.71	2.56
29	4.18	3.33	2.93	2.70	2.54

Denominator df (left-margin label)

The shaded value is the critical value of F for the one-way ANOVA performed on data for the effect of placebo pills on blood pressure.

software) and the magnitude of the critical value of F for at $\alpha = .05$ varies with the degrees of freedom, both those for the treatments and error. As we anticipate that F will increase if the null hypothesis is false, the numerator is *always* the treatment MS and the denominator the error MS. Unlike the t-test we do not have a choice of one-tailed or two-tailed tests depending on the null hypothesis being tested, the null hypothesis is already determined for us.

The pertinent section of an F table for $\alpha = .05$ below shows that with our degrees of freedom (3 for the numerator or treatment MS, and 28 for the denominator or error MS) the critical value of F is 2.95 (Table 6.5). So in this case to reject the null hypothesis and conclude that the treatments had an effect we require approximately three times the variance associated with the treatments (fixed factor) compared with the error. Our calculated F ratio is much bigger than this at 24.64 so $p < .05$ and we would reject the null hypothesis that every sample is drawn from a population(s) with the same population mean. Note that this does not tells us whether the samples are drawn from two or more populations with the same population mean or whether they are all samples from the same population with a single population mean. Note also that the smaller the sample size, and therefore error degrees of freedom, the larger the critical value of F. For smaller samples the effect of the treatments on the variance has to be much larger to attain statistical significance. We could report this F ratio as $F_{3,28} = 24.64$, where by convention the subscripted numbers are the treatment and error degrees of freedom respectively.

The results of our ANOVA tell us that there is an effect of treatment, at least one of the placebos alters blood pressure, but it does not tell us which particular treatments are different from each other. To find this out we need to compare the groups using a comparison test. There are several tests that can be used in conjunction with an ANOVA and which we use depends partly on the comparisons we wish to make. We will briefly describe how to interpret two commonly used tests, Tukey's and Dunnett's. Both are based on the *t*-test, but unlike Bonferroni's correction, these tests correct for multiple comparisons by adjusting α in a less conservative way.

6.11.1 Tukey's honest significant difference test

Tukey's post hoc test can compare all possible pairs of means, where the overall (family) value for α is .05 and the value of α for each comparison is determined by the number of treatments, k, the error degrees of freedom, v, and a test statistic called q. As long as the test assumptions are met (homogeneity of variances, independence and normality) it is a robust test that maintains α at intended values.

The output from Minitab for a one-way ANOVA is reported in Box 6.2. The first part is the ANOVA itself, which is what we calculated above. Below the ANOVA results are the sample descriptive statistics and 95% confidence intervals. It looks like the white pills and blue pills might have lowered the systolic blood pressure as the means are lower than no treatment and the confidence intervals don't overlap, but Tukey's test below will tell us for sure. Recall that the 95% confidence intervals estimate the location of the population mean. The individual value of α, that is the significance level for each comparison, is .0108. This is less conservative than Bonferroni's correction where it would be .0083 for six comparisons. The critical value of the test statistic, q, is 3.86 and for each comparison calculated q must exceed this for the difference to be significant at the .05 level. To interpret the table of intervals at the bottom, we need to look at each pair of numbers for a given comparison and if the pair includes zero there is no difference between those treatments, whereas if zero is not included there is a difference. So for no pills (none) versus blue pills the interval is −23.069 to −9.181. Both numbers are negative so they don't span zero. We can conclude that blue placebo pills lower blood pressure in patients when compared with the blood pressure of patients given no pills. For the comparison no pills versus red pills, however, the values are −11.194 to 2.694. As one number is negative and the other is positive, this interval spans zero so there is no difference in the systolic blood pressure of patients given no pills compared with those given red placebo pills. Now try to work out for yourself whether there are differences for the other comparisons. You should find that the results are as follows:

Blue versus red – difference, $p < .05$

Blue versus white – difference, $p < .05$

None versus white – difference, $p < .05$

Red versus white – difference, $p < .05$

Box 6.2   *Minitab output for a one-way analysis of variance with Tukey's pairwise comparisons*

```
Analysis of variance for placebo

Source    DF        SS       MS        F       P
Placebo    3    1913.6    637.9    24.64   0.000
Error     28     724.9     25.9
Total     31    2638.5


                                  Individual 95% CIs for mean
                                  Based on pooled StDev
Level      N     Mean    StDev    ------+---------+---------+---------+
blue       8   127.38     5.24    (---*----)
none       8   143.50     4.24                        (---*----)
red        8   147.75     4.20                           (----*---)
white      8   136.25     6.36               (---*----)
                                  ------+---------+---------+---------+
Pooled StDev = 5.09               128.0     136.0     144.0     152.0


   Tukey's pairwise comparisons

   Family error rate = 0.0500

   Individual error rate = 0.0108

   Critical value = 3.86

   Intervals for (column level mean) - (row level mean)

                  blue       none       red

       none    -23.069
                -9.181

       red     -27.319   -11.194
               -13.431     2.694

       white   -15.819     0.306     4.556
                -1.931    14.194    18.444
```

6.11.2   Dunnett's test

Suppose we aren't interested in the pairwise comparisons produced by Tukey's test but only in whether placebo pills affect blood pressure compared with a control, in this case no pills. In this instance we could use a Dunnett's test, in which a treatment is nominated as the control and each of the other treatments is compared with it. This reduces the number of comparisons when compared with Tukey's test. For our placebo data, there are four groups and so three comparisons. The Minitab output for a Dunnett's test of placebo pills data is reported in Box 6.3. The first part of the output is identical to the ANOVA we performed earlier so we need to concentrate on the second part of the output below it, Dunnett's comparisons with a control. Again, with the significance level, α, at .0193 for individual comparisons, Dunnett's test is less conservative than multiple *t*-tests with Bonferroni's correction where it would be .0167 when three tests are required. The critical value of the test statistic (also q) is 2.48 and for each comparison calculated q must exceed this for the difference to be significant at the .05 level. The

---

Box 6.3   Minitab output for a one-way analysis of variance with Dunnett's comparisons with a control

```
Analysis of variance for placebo

Source    DF      SS      MS      F       P
Placebo    3   1913.6   637.9   24.64   0.000
Error     28    724.9    25.9
Total     31   2638.5


                              Individual 95% CIs For Mean
                              Based on Pooled StDev
Level    N    Mean   StDev   ------+---------+---------+---------+
Blue     8  127.38    5.24   (---*----)
None     8  143.50    4.24                   (---*----)
Red      8  147.75    4.20                       (----*---)
White    8  136.25    6.36            (---*----)
                              ------+---------+---------+---------+
Pooled StDev = 5.09                 128.0   136.0   144.0   152.0


Dunnett's comparisons with a control

Family error rate = 0.0500

Individual error rate = 0.0193

Critical value = 2.48

Control = level of treatment(none)

Intervals for treatment mean minus control mean

Level     Lower    Center    Upper   ---+---------+---------+---------+----
White   -13.567   -7.250   -0.933            (------*-----)
Red      -2.067    4.250   10.567                    (-----*------)
Blue    -22.442  -16.125   -9.808    (-----*-----)
                                     ---+---------+---------+---------+---
                                      -20      -10       0       10
```

diagram of 95% confidence intervals for Dunnett's test differs from that for the ANOVA in that the comparison is now for a *difference* in means between each of the placebo pill groups and no pill group, so the scale runs from −20 to 10. If we subtract the mean of the control group 'none' from the mean of the 'red', the difference is $147.75 - 143.50 = 4.25$. The mean for this comparison is represented by the asterisk at 4.25 on the *x*-axis. Although the systolic blood pressure is a little higher in patients given red placebo pills, the difference is not statistically significant because the confidence interval for the difference in means spans zero. For both of the other comparisons, no pills versus white pills and no pills versus blue pills, there is a statistically significant decrease in the means – the asterisks are at negative values on the *x*-axis and the confidence intervals do not overlap zero. These results mirror those we reported for Tukey's test.

One-way ANOVA results summary: A difference was observed in the mean systolic blood pressure of patients given different coloured placebo pills or no pills ($F_{3,28} = 24.64, p < .001$).

For Tukey's comparisons:
Tukey's pairwise comparisons revealed that the mean systolic blood pressure of patients administered red pills (147.75 mmHg) did not differ from that of untreated patients (143.50 mmHg). The administration of white or blue placebo pills, however, resulted in mean systolic blood pressures of 136.25 mmHg and 127.38 mmHg respectively which were significantly lower than systolic pressures of patients given no pills or red pills (Tukey's family error rate = .0500, individual error rate = .0108).

For Dunnet's comparisons:
Dunnet's test to compare pills versus the no pills control patients revealed that the mean systolic blood pressure of patients given red pills was 147.75 mmHg but did not differ from that of controls which was 143.50 mmHg (95% CI for the difference −2.067, 10.567). The mean systolic blood pressure of patients given either white or blue pills, however, was lower than that of patients given no pills (mean$_{white}$ 136.25 mmHg, 95% CI for the difference −13.567, −0.933, mean$_{blue}$ 127.38 mmHg, 95% CI for the difference −22.442, −9.808). Dunnet's family error rate = 0.0500; individual error rate = 0.0193.

6.11.3 Accounting for identifiable sources of error in one-way ANOVA: nested design

Let's suppose we are investigating oxidative stress and we measure superoxide levels in mice of different ages with a luminescence assay. This simply involves a chemical reaction that produces light, the more superoxide, the more light units counted by the detector in the plate reader.

We have four different age groups of mice; each group comprises three mice housed together in one cage from when they were weaned. We take one piece of liver tissue from each mouse for the assay. Is this an appropriate experimental design to assess the effect of age on superoxide levels of mice? Not really, after all it could be that any differences in superoxide observed between treatment (i.e. age) groups could be due to random variation perhaps related to the cage (more food, nicer occupants, warmer, etc.). We have selected cages (necessary, or the mice escape) but apart from the requirement that they retain the animals, we have not chosen them for any particular property. To avoid this uncertainty a better design is to use replicate cages for each treatment group – in the example below we have four cages of three mice for each age group – so that we can assess whether the variation *between* age groups is really different from the variation *between* the cages, and likewise we can assess whether the variation *between* cages is no different from the variation *within* cages.

This design is called a *nested hierarchical design* and requires a rather different one-way ANOVA from the previous section. Each group has the factor 'cages' nested within it, such that it is subordinate to the age group. Hierarchical and subordinate in this context indicate that cages cannot be randomly assigned to a treatment. A cage contains mice of a particular age, so a cage of old mice cannot be assigned to the treatment 'juvenile mice' (or else we'd have discovered how to reverse the aging process). Regrettably, an old mouse remains an old mouse and must be assigned to the treatment 'old mice'. We can say that 'cage' is subordinate to age, that is, there is a hierarchical structure to the factors; age dictates which cage is allocated to which treatment. Furthermore, the subordinate factor 'cage' is not the usual fixed factor we saw in the previous section, but rather a random factor. By this we mean that we, as the experimenters, have no real control over any variation associated with the cages. This is in contrast to the fixed factor 'age', which has been determined by us. In biology, this arrangement, a random factor nested within a fixed factor is common, but a second factor that is fixed does sometimes occur (e.g. gender, we could also have males and females within a cage). This affects the ANOVA as we shall see.

The data for the following example are in Appendix A and the experimental design is illustrated in Figure 6.7. Four age groups were selected: juveniles (4 weeks), young adults (10 weeks), mature adults (26 weeks) and old adults (78 weeks), so we have 3 degrees of freedom for the fixed factor age (= number of treatment levels − 1). We also have four cages nested within each age group (treatment level). The conventional way to represent a nested random factor is write it next to the fixed factor to which it is subordinate enclosed within brackets so here either Cage(Age) or (Age)Cage. The degrees of freedom for the Cage(Age) would be (number of cages per

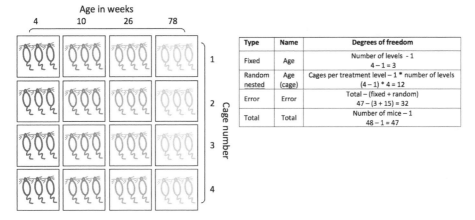

Figure 6.7 A nested ANOVA design. On the left is a schematic representation of the experimental design. A cage of young mice cannot be swapped for a cage of old mice so we say that the subordinate random factor 'cage' is nested within the fixed factor 'age'. The table on the right details how the degrees of freedom for this design are calculated.

treatment group − 1) multiplied by the number of treatment groups, that is, $(4 − 1) * 4 = 3 * 4 = 12$. As the total number was 3 mice $*$ 4 cages $*$ 4 ages we have 48 mice, so $n − 1 = 47$ total degrees of freedom. Hence by subtraction the error degrees of freedom must be the total minus the fixed and random, that is, $47 − (12 + 3) = 32$.

So what is different about this ANOVA? We have calculated the MS in the usual way (SS/df) but the way in which the null hypothesis is tested differs (Box 6.4). As we wish to know whether the treatment has added any extra variation in addition to that of the cages we test for significance by an F ratio computed as $MS_{stressor}/MS_{cage(stressor)}$, not as you might first think using the error MS as the denominator.* So here $F = 343.7 / 118.49 = 2.90$. The error MS is used as the denominator to calculate the F value for the factor Cage, if it were of any interest $F = 118.49 / 93.43 = 1.27$.

Box 6.4 *Minitab output for a nested analysis of variance for four cages each with three mice nested within each of four age groups*

```
Analysis of variance table

Source      DF    SS        MS       F       P
Age         3     1030.10   343.37   2.90    0.079
Cage(age)   12    1421.89   118.49   1.27    0.284
Error       32    2989.88   93.43
Total       47    5441.9
```

---

*Rather confusingly if the nested factor is a fixed factor, then the F test is indeed $MS_{stressor}/MS_{error}$, but this is comparatively rare in biology.

As the analysis reveals, when the random factor Cage is accounted for there was no significant effect of age on superoxide activity of the liver ($p = .079$). What would we have found if we had simply ignored the fact that mice were in different cages? A simple one-way ANOVA with four treatment groups, each of 12 mice, would suffice (Box 6.5). And in this case we would conclude, wrongly for this design, that age had a substantial effect as $p = .025$, much lower than .05. Obviously, some of the variation between age groups was in fact originating from the effect of caging.

Box 6.5   Minitab output for a one-way analysis of variance for the same data as those used in the nested design. Four age groups each with 12 mice

```
Analysis of variance table

Source   DF   SS        MS       F      P
Age      3    1030.10   343.37   3.42   0.025
Error    44   4411.8    100.3
Total    47   5441.9
```

Nested ANOVA results summary: No difference in liver superoxide levels was observed in mice of different ages with Cage as a nested factor ($F_{3,32} = 2.9, p = .079$).

6.12   Two-way ANOVA

We have examined how the error sum of squares can be partitioned to account for unknown random factors in a simple one-way ANOVA or for where we suspect a source random error by using a nested design, but we can also subdivide the treatment sum of squares when we have a second factor of interest that we would like to test. Let us take the example we used in our first ANOVA, the effect of placebo pills on blood pressure. In that example we administered the treatments for 2 weeks, but suppose that we also wanted to know whether the duration of treatment might have an effect, we could give half the subjects in each group treatment for 2 weeks, and half of them treatment for 4 weeks. We now have two factors: treatment (no pills or pills) and duration (2 or 4 weeks). In this instance we could use a two-way ANOVA, meaning that there are two factors of interest or fixed factors (note that this differs from the nested design, where there is actually only one factor of interest, the fixed factor).

When multiple comparisons are requested, the Minitab output is divided into three parts: the overall ANOVA, the comparisons for the first factor and the comparisons for the second factor (Box 6.6A–C). There is an effect of the treatment ($F = 23.74$, $p = < .001$) but no effect of duration of treatment ($F = 0.14$, $p = .711$). Note that we have introduced another term into the model: treatment * duration. This is called the *interaction* between the fixed factors. We can omit it from our model, but it is useful to know about any interactions because sometimes they achieve statistical significance and tell us something else about our data. In the example below the interaction is not statistically significant ($F = 0.95$, $p = .433$), but suppose that our second factor were say, gender rather than time, and we found a significant interaction. This might occur if, for example, different colour placebo pills had strongly opposite effects on systolic blood pressure in men compared with women. The interaction between the fixed factors would be of interest here because it would be useful to know that giving placebo pills of a certain colour increases systolic blood pressure in women and decreases it in men.

Box 6.6(A) *Two-way ANOVA table for the effect of placebo pill colour and duration of treatment on systolic blood pressure*

```
General linear model: systolic versus treatment, duration

Factor              Type    Levels   Values
Treatment           fixed      4       Blue     None    Red    White
Duration            fixed      2        2        4


Analysis of variance for systolic, using adjusted SS for tests

Source             DF    Seq SS    Adj SS   Adj MS      F       P
Treatment           3   1913.59   1913.59   637.86   23.74   0.000
Time                1      3.78      3.78     3.78    0.14    0.711
Treatment*duration  3     76.34     76.34    25.45    0.95    0.433
Error              24    644.75    644.75    26.86
Total              31   2638.47
```

The second part of the output contains Tukey's pairwise comparisons for all levels of treatment, first the 95% confidence intervals on a scale for the difference between means, then tables containing the difference between means, its standard error, t and p. For the comparison of blue placebo pills versus each of the other groups, the confidence intervals for the difference do not span zero and all three p-values in the table are $< .05$, so the effect of blue placebo pills on systolic blood pressure differs from all other pill colours or no pills. Compare the results of the other comparisons with the one-way ANOVA performed for treatment earlier. You should find that the same differences between groups are revealed.

The final part of the output contains Tukey's pairwise comparisons for both levels of treatment duration, first the 95% confidence interval on a scale for

Box 6.6(B)   *Tukey's pairwise comparisons for the fixed factor treatment (i.e. pill colour or no pills)*

```
Tukey 95.0% simultaneous confidence intervals
Response variable systolic
All pairwise comparisons among levels of treatment

Treatment = blue subtracted from:

Treatment   Lower   Centre  Upper   ---+---------+---------+---------+---
None        8.978   16.125  23.27                   (----*----)
Red        13.228   20.375  27.52                    (----*----)
White       1.728    8.875  16.02            (----*---)
                                    ---+---------+---------+---------+---
                                    -15        0        15        30


Treatment = none subtracted from:

Treatment   Lower   Centre   Upper    ---+---------+---------+---------+---
Red         -2.90    4.250   11.3968      (----*----)
White      -14.40   -7.250   -0.1032  (----*---)
                                      ---+---------+---------+---------+---
                                      -15        0        15        30

Treatment = red subtracted from:

Treatment   Lower   Centre   Upper    ---+---------+---------+---------+---
White      -18.65  -11.50    -4.353   (---*----)
                                      ---+---------+---------+---------+---
                                      -15        0        15        30

    Tukey simultaneous tests
    Response variable systolic
    All pairwise comparisons among levels of treatment

    Treatment = blue subtracted from:

    Level        Difference     SE of             Adjusted
    treatment    of means     difference  T-Value  P-Value
    None           16.125       2.592      6.222    0.0000
    Red            20.375       2.592      7.862    0.0000
    White           8.875       2.592      3.425    0.0111


    Treatment = none subtracted from:

    Level        Difference     SE of             Adjusted
    treatment    of means     difference  T-Value  P-Value
    Red             4.250       2.592      1.640    0.3763
    White          -7.250       2.592     -2.798    0.0460


    Treatment = red subtracted from:

    Level        Difference     SE of             Adjusted
    treatment    of means     difference  T-Value  P-Value
    White         -11.50        2.592     -4.437    0.0009
```

Box 6.6(C) *Tukey's pairwise comparisons for the fixed factor treatment duration*

```
Tukey 95.0% simultaneous confidence intervals
Response variable systolic
All pairwise comparisons among levels of duration


Duration = 2 subtracted from:

Duration    Lower    Centre    Upper    -------+---------+---------+-------
2          -4.470   -0.6875   3.095    (-------------*-------------)
                                        -------+---------+---------+-------
                                            -2.5      0.0      2.5


   Tukey simultaneous tests
   Response variable systolic
   All pairwise comparisons among levels of duration


   Duration = 1 subtracted from:

   Level       Difference       SE of                   Adjusted
   duration     of means     difference    T-Value      P-Value
   2             -0.6875        1.833       -0.3752      0.7108
```

---

the difference between treatment duration means, then tables containing the value of the difference, its standard error, t and p. The overall ANOVA in the first part informed us that there was no difference between patients taking pills for 2 weeks and those taking pills for 4 weeks so we would not expect the Tukey's test to reveal differences and indeed it does not.

Two-way ANOVA results summary: A difference was observed in the mean systolic blood pressure of patients given different coloured placebo pills or no pills ($F_{3,24} = 23.74$, $p < .001$). However, there was no effect of duration of treatment on mean systolic blood pressure ($F_{1,24} = 0.14$, $p = .711$). No interaction was observed between treatment and duration of treatment ($F_{3,24} = 0.95$, $p = .433$).

Tukey's pairwise comparisons revealed that the mean systolic blood pressure of patients administered red pills was 147.75 mmHg and did not differ from that of untreated patients, which was 143.50 mmHg (95% CI for the difference −2.90, 11.40). The administration of white placebo pills, however, resulted in a lower mean systolic blood pressure of 136.25 mmHg relative to no pills (95% CI for the difference −14.40, −0.10, $p = .046$) or red pills (95% CI for the difference −18.65, −4.35, $p = .0009$). Patients administered blue placebo pills exhibited the lowest mean systolic blood pressure, 127.38 mmHg, which was significantly different from all other treatment groups, $p < .05$.

6.12.1   Accounting for identifiable sources of error using a two-way ANOVA: randomised complete block design

In the last section we examined how to incorporate a second factor for nested, hierarchical designs, but there are situations where the second factor is random but not nested, that is, it is not subordinate to another factor. Because an ANOVA works by partitioning the total variation in the data into that due to treatment and that due to error, if we can quantify a *known* random factor, we can include it in our analysis to increase the likelihood of detecting a treatment effect. Let's look at example of an experiment in which this might be the case.

Imagine that we want to investigate the effects of three compounds A, B and C dissolved in culture medium on cell migration using a time-lapse video microscopy system that allows us to track the movement of cells cultured in a six-well plate. We plan to have five replicate wells of cells for each experimental treatment including the negative control treatment, labelled D, which is culture medium alone. In total we would have 20 wells of cells; five replicates for each of four treatments. We record the cells overnight and then use some software to work out the distance in micrometres that the cells have moved, our response variable. The trouble is that we have only one instrument and booking time slots for its use has to fit in with other users. We need to design the experiment so that we can take into account the fact that we will have to stagger treating the cells and collecting the migration data over several nights. But by running the experiment on different nights we are introducing an additional random factor as conditions on one night may not be equivalent to those on another night. As we **know** that this might be a potential source of error variation, we can work it into our ANOVA model. This is called a *randomised complete block design* (or sometimes simply *randomised block design*).

From a practical point of view, there are two ways we could design the experiment: (1) run all five replicates of one treatment on a single night or (2) run one replicate of each treatment on each night. If we wish to include our known random factor (on which night a plate of cells is run) into the statistical model, which of these two options do you think would be the better experimental design? The only way we can address the potential effect of different nights is to use the second experimental design (Figure 6.8, colour version is located in colour plate section). If we use the first design we would not be able to tell whether treatment differences were simply due to the fact that we had run each one on a different night, so it would be a poor choice of design. With the second design, all treatments are run each night, so if there are effects of different nights, we assume that they would affect all treatments equally and the model would be able to detect them. We call the potential random

Block design (one plate of cells per night)

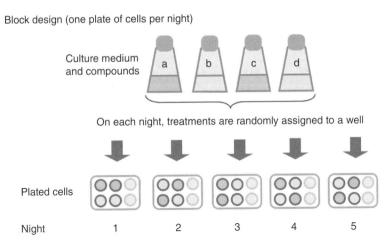

Figure 6.8   Randomised complete block ANOVA design. Any of the four treatments, a–d, may be applied randomly to wells in one of the five plates and any plate may be run on any of five nights (blocks). The ANOVA partitions the SS with 'blocks' as an identified potential source of variation. (For a colour version, see the colour plate section.)

factor a *block*, and as we shall run the experiment on five nights, the factor has five levels.

It is also essential that the cells are assigned randomly to the wells on each night so that there is no positional effect due to the wells. Statisticians refer to the factors being *crossed*, that is, every level of one factor is found in combination with every level of the other factor (in our case all four treatments are present on each night), which is not the case for hierarchical designs.

The fictitious raw data for our randomised complete block experiment are located in Appendix A. We have run two different ANOVA models with the data in order to compare what happens when blocks are excluded and included. The first set of results (no blocks) uses the same model as the ANOVA we ran for the first placebo pills experiment above. The test statistic F is only 2.67, which is below the critical level of 3.2389 for 3 numerator and 16 denominator degrees of freedom with α at .05, so there is apparently no statistically significant difference between the four treatment means (Box 6.7). According to this test, only a third of the variation is attributable to treatment and almost two thirds to random error (treatment SS of 125,380 is about 33% of the total SS and error SS of 250,400 is about 67% of the total SS).

Now let's see what happens if we account for the known random factor of different nights by adding it as blocks into the design (Box 6.8). The total degrees of freedom are unchanged as that is always number of observations − 1 so here it remains at 19 (we still have 20 wells of cells). The treatment degrees of freedom are also unchanged at number of treatments − 1 = 3. The

---

Box 6.7   Minitab output of the ANOVA table for migration distance versus treatment without blocking, that is, a one-way design

```
ANOVA: migration distance versus treatment (no blocks)

Factor       Type    Levels   Values
Treatment    Fixed      4     a, b, c, d

       Analysis of variance for migration distance

       Source      DF      SS      MS      F       P
       Treatment    3   125380   41793   2.67   0.083
       Error       16   250400   15650
       Total       19   375780
```

---

difference is that the blocks now account for 4 degrees of freedom, as it has five levels (nights) and we subtract 1. So the error degrees of freedom are now reduced to 12, which has the effect of reducing the error mean squares to only 3939. This in turn means that as we now divide the treatment means squares, 41,793, by a smaller number for the error means squares, 3939, our *F* ratio is much bigger at 10.61. This now exceeds the critical value, which for 3 numerator and 12 denominator degrees of freedom with α at .05, is 3.4903. By accounting for the variation due to different nights we now have a smaller contribution to the variation from unknown random factors, and we see that there is a statistically significant effect of treatment on cell migration distance ($p = .001$).

Box 6.8   Minitab output of the ANOVA table for migration distance versus treatment with blocking, that is, a two-way design

```
ANOVA: migration distance versus treatment, blocks

Factor       Type    Levels   Values
Treatment    Fixed      4     a, b, c, d
Blocks       Random     5     1, 2, 3, 4, 5

       Analysis of variance for migration distance

       Source      DF      SS      MS       F       P
       Treatment    3   125380   41793   10.61   0.001
       Blocks       4   203130   50783
       Error       12    47270    3939
       Total       19   375780
```

You may at this point think to yourself 'how does this differ from nested ANOVA?'. The answer is that in the randomised block example, the four treatments can be assigned to any of the wells and a particular plate can be prepared and the experiment can be run on any of the nights. There is no hierarchy so no factor is subservient to another. In the nested example, an old mouse cannot be transformed into a young mouse. Therefore the factor Cage is subservient to the factor Age: a cage of mice cannot be assigned to any age group randomly, the treatment to which it is assigned, young or old, is dictated by the age of the mice it contains. So a nested design is appropriate when there is a hierarchical relationship between factors and one is subservient to another (Figure 6.7). A randomised block design is relevant when there is no hierarchy between factors.

> **Two-way ANOVA randomised block design results summary:**
> There was a statistically significant difference in the migration distance of cells treated with either culture medium alone or three different compounds ($F_{3,12} = 10.61$, $p = .001$).

6.12.2 Repeated measures ANOVA

Suppose that we have a variable that we measure in the same individuals on different occasions (levels). Similar to data used for a paired t-test, the measurements at each level are not independent of each other. One of the assumptions of the ANOVA models described above is that the measurements between samples are independent of each other, so we cannot use any of these if we have this type of experimental design. We would need to use a *repeated measures* ANOVA. This model takes into account of the dependence between the groups and is therefore appropriate for these data. This type of design is not unusual in crossover clinical trials where the same subject receives all treatments. The repeated measure ANOVA can partition the variation from differences between individuals (the test is sometimes called a *within-subjects* ANOVA). The total sum of squares is comprised of treatment sum of squares and error sum of squares as in other models, but there is an additional factor, subjects sum of squares, which is the variation due to individual differences.

6.13 Summary

Where data are normally distributed, randomly drawn from the population and variances are equal, parametric tests that compare means may be used. For two groups, t-tests are appropriate. Where variances between *two* groups are not equal, a modification of the *t*-test, called Welch's *t*-test can be used. Where more than two groups are to be compared, an ANOVA is appropriate. The simplest ANOVA model partitions the variance into that attributable to fixed factor and that attributable to random factors (known as the error or residual). More complex models can account for repeated measures or subdivide the variance (e.g. blocking, nesting, two-way) to either account for known sources of error or where there is more than one factor of interest.

References

Blackwell, B., Bloomfield, S.S. and Buncher, C.R. (1972) Demonstration to medical students of placebo responses and non-drug factors. *The Lancet*, 299(7763):1279–1282.

Efron, B. (1978) Regression and ANOVA with zero-one data: measures of residual variation. *Journal of the American Statistical Association*, 73(361):113–121.

Health and Social Care Information Centre (2011) Health Survey for England, 2008. Available at www.hscic.gov.uk (accessed 30 June 2013).

Joiner, B.L. (1975) Living histograms. *International Statistical Review*, 43(3):339–340.

Lango Allen *et al.* 2010. Hundreds of variants clustered in genomic loci and biological pathways affect human height. *Nature*, 467(7317):832–838.

Lim, T-S. and Loh, W-Y. (1996) A comparison of tests of equality of variances. *Computational Statistics and Data Analysis*, 22:287–301.

McDowell, M., Fryar, C.D., Ogden, C.L. and Flegal, K.M. (2008) Anthropometric reference data for children and adults: United States, 2003–2006. *National Health Statistics Reports*, 10:15

Razali, N.M. and Wah, W.B. (2011) Power comparisons of Shapiro-Wilk, Kolmogorov-Smirnov, Lilliefors and Anderson-Darling tests. *Journal of Statistical Modeling and Analytics*, 2(1):21–33.

Waber, R.L., Shiv, B., Carmon, Z. and Ariely, D. (2008) Commercial features of placebo and therapeutic efficacy. *Journal of the American Medical Association*, 299(9):1016–1017.

Plate 4.20 Numbers represented on a colour scale from light green to dark green. The colours could represent any range of numbers as long as a scale bar is clearly reported.

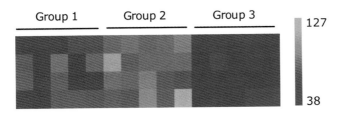

Plate 4.21 Heat map representing the data in Table 4.2. These colour graphs are commonly used to summarise genomic data such as gene expression changes. Green to red is very popular but may be difficult for colour-blind readers to distinguish differences so blue to orange or yellow to red are better.

Starting Out in Statistics: An Introduction for Students of Human Health, Disease, and Psychology
First Edition. Patricia de Winter and Peter M. B. Cahusac.
© 2014 John Wiley & Sons, Ltd. Published 2014 by John Wiley & Sons, Ltd.
Companion Website: www.wiley.com/go/deWinter/startingstatistics

Block design (one plate of cells per night)

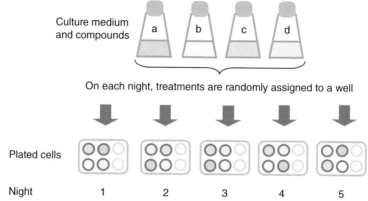

Plate 6.8 Randomised complete block ANOVA design. Any of the four treatments, a–d, may be applied randomly to wells in one of the five plates and any plate may be run on any of five nights (blocks). The ANOVA partitions the SS with 'blocks' as an identified potential source of variation.

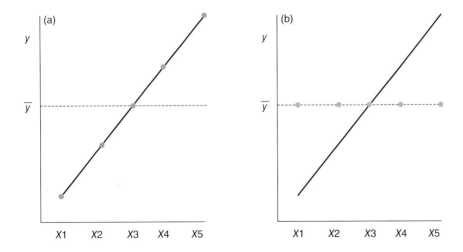

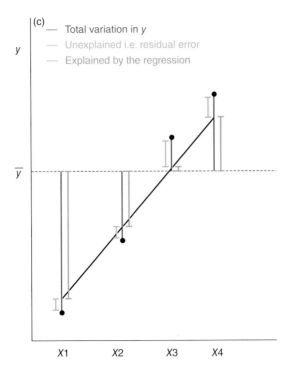

Plate 7.4 Sources of variation in linear regression. (a) If all the values of y lie perfectly on the regression line, then all the variability in y is due to the regression and there is no variation due to residual error. (b) If all the values lie on the grand mean of y, ȳ, represented by the dotted horizontal line, all the variation in y is due to residual error and there is no effect of the explanatory variable x. (c) In practice, the total variation in y comprises both regression and residual error. The total variation is represented by the blue lines and is the distance of the data points from ȳ. The variation due to residual error is represented by the distance of the data points to the black regression line (orange lines) and when subtracted from the total, gives the portion of the variation that is explained by the regression (green lines).

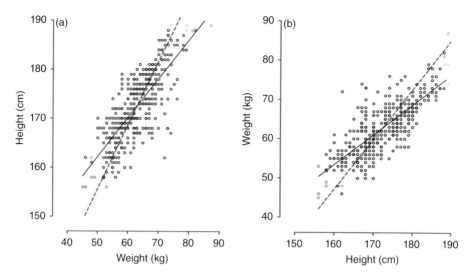

Plate 7.8 Regression for female swimmer height and weight plotted with (a) height as the dependent variable and (b) weight as the dependent variable as in the original analysis. The regression line for each plot is the solid black line. The dotted grey line is the regression line for the alternative plot, that is, in plot (a) it is the regression line for plot (b) and vice versa. To aid orientation selected data points have been coloured the same in each plot. Solid circles are those bisected by a regression line. It is clear that the two solid black regression lines do not run through the same points in each plot and hence differ in their slopes.

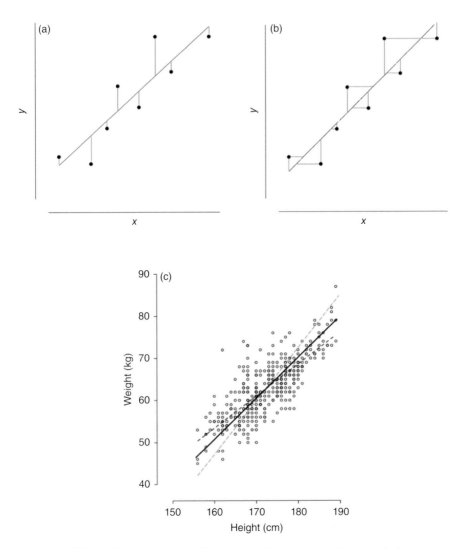

Plate 7.9 Difference between ordinary linear regression and reduced major axis linear regression. (a) For OLS the line is fitted to reduce the error in *y*. (b) For RMA the line is fitted to account for error in both *y* and *x*. (c) The RMA regression line (solid black) has been fitted to the female swimmer data from Figure 7.6. The dotted lines are the OLS regression lines for when height is plotted on *x* (blue) or weight is plotted on *x* (orange).

Table 10.2 Demonstration of the bootstrap resampling method

Original	Bootstrap 1	Bootstrap 2	Original	Bootstrap 1	Bootstrap 2
99	**99**	99	112	113	110
103	**103**	99	113	113	112
104	104	103	113	113	112
104	104	103	113	113	**113**
104	106	103	113	113	**113**
105	106	**104**	113	113	**113**
106	106	**104**	113	113	**113**
107	107	**104**	113	113	**113**
107	108	**105**	114	113	**113**
108	108	106	115	113	**113**
108	108	106	115	113	114
108	108	**107**	115	113	114
109	**109**	**107**	117	115	115
109	**109**	108	117	115	115
110	110	108	118	118	115
110	**111**	108	118	118	115
110	**111**	108	120	118	118
111	**111**	108	126	**128**	118
111	**112**	108	128	**128**	118
111	**112**	109	130	130	126
112	**112**	109	130	130	126
112	**112**	109	135	130	128
112	**112**	109	135	130	130
112	**112**	109	142	135	135
112	113	110			

The bootstrap samples were randomly resampled with replacement from the column of original data, which are resting systolic blood pressures ranked low to high $n = 49$, $\bar{x} = 113.9$ mmHg. Values in blue represent numbers sampled the same number of times in the bootstrap samples as they occurred in the original data. Values in green appear more times in the bootstrap sample than the original, for example, there is one value of 106 in the original sample but it has been sampled three times in bootstrap sample 1 and twice in bootstrap 2. Values in orange appear fewer times in the bootstrap samples than in the original, that is, 110 appears three times in the original sample but only once in bootstrap sample 1 and twice in bootstrap sample 2. Finally, values in the original samples may be omitted from a bootstrap sample, for example, 105 does not appear in bootstrap 1 and 111 is also missing in bootstrap 2.

7

Relationships between Variables: Regression and Correlation – 'In Relationships ... Concentrate only on what is most Significant and Important'

7.1 Aims

The title is a quotation from Danish Philosopher Soren Kierkegaard. 'It seems essential, in relationships and all tasks, that we concentrate only on what is most significant and important'. This chapter deals with regression and correlation, statistical tests that students often mistake for each other as they both involve determining whether there is a relationship between two continuous variables, where each case has two measurements. We will clarify the differences between the two by explaining that regression is primarily predictive, fitting a line to the data, whereas correlation quantifies the strength of a linear relationship between two variables. We will first describe linear regression, because this progresses naturally from ANOVA in the previous chapter. Graphical illustrations will be used to explain the principal of regression – determining whether the slope of the fitted regression line differs from

Starting Out in Statistics: An Introduction for Students of Human Health, Disease, and Psychology
First Edition. Patricia de Winter and Peter M. B. Cahusac.
© 2014 John Wiley & Sons, Ltd. Published 2014 by John Wiley & Sons, Ltd.
Companion Website: www.wiley.com/go/deWinter/startingstatistics

a slope of zero (a horizontal line), given that there is variability in y at each value of x. We will also discuss situations where x and y are random variables. For correlation, we will first explain the use of Pearson's correlation analysis and then its non-parametric alternative, Spearman's rank test, for data which do not satisfy certain assumptions for Pearson's correlation coefficient.

7.2 Linear regression

In the previous chapter we learned that if we wish to compare multiple means to test a hypothesis that $\mu_1 = \mu_2 = \mu_3 = \cdots = \mu_n$, we can use an ANOVA, which partitions the total variance into that attributable to treatment, usually fixed factor(s), and that attributable to random error. Regression analysis is similar to ANOVA, in that we partition the sums of squares into variation resulting from the regression and that resulting from residual error. A linear regression is appropriate when we wish to quantify if there is a linear relationship between two continuous variables, where the values of the independent variable on the x-axis accounts for some of the variability in the dependent variable on the y-axis. It is useful if we would like to use paired data to which we have fitted a regression line in order to predict an unknown value of one variable when we know its value for the other axis. A common example of this is a calibration (standard) curve for a biochemical assay, such as one that quantifies the amount of a substance, such as protein, in samples from its absorbance measured in a spectrophotometer. The absorbances (dependent variable) of a series of standards (samples of known protein concentration) are measured. The data for the standards are graphed and a linear regression line is fitted. This graph is the calibration curve and it is then used to determine (predict) the concentration of protein in samples of unknown concentration using their measured absorbance values. Confidence intervals may also be determined so that the accuracy of the predicted measurements can be evaluated.

Linear regression can be used for prediction in many situations, once it is established that a statistically significant relationship between two variables exists. For example, a surgeon may wish to know if the volume of blood lost during routine planned surgery is linearly related to preoperative blood pressure in order to predict which patients are at risk of greater blood loss. If a relationship between the variables exists, preventative action may be taken, for example, reduce the patient's blood pressure before surgery. Note that if a significant relationship were observed this does not mean that high blood pressure *causes* blood loss, the causative factor may be something completely different.

The linear regression is described by the mathematical equation for a straight line, $y = mx + c$, where m is the slope of the line and c is the intercept of the line with the y-axis when x is 0. In Figure 7.1, the intercept, c, has a value

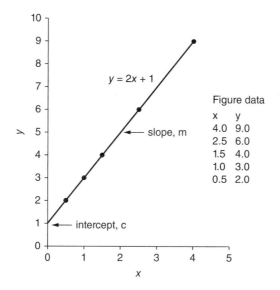

Figure 7.1 Plot of the mathematical relationship between the numbers in the table to the right. The points fall on a straight line for which the equation is $y = 2x + 1$. Any value of x may be multiplied by 2 and added to 1 to give the value of y. The line intercepts the y-axis at 1 when x is 0, and has a slope of +2.

of 1 and the slope has a value of 2. The intercept may be a positive or negative value. Likewise the slope may also be either positive or negative, but for both the larger the absolute value of the slope, the steeper the line.

It would be very nice and easy if the relationship between two variables were as perfectly straight as that of the numbers in Figure 7.1, but we should know by now that there is always variability associated with measurements, which is why we need to use a linear regression model to account for this variability. The variable that is plotted on the x-axis of a linear regression is analogous to a fixed factor in an ANOVA in that we wish to know how much of the variation in the data on the y-axis it explains. We anticipate that if increases in the independent variable x, change the values of the dependent variable y, proportionally there may be a linear relationship between the two variables. We can then suggest that the value of the independent variable may be used to predict the value of the dependent variable or vice versa. Note how this differs from the fixed factor 'placebo pills' in the one-way ANOVA example of the previous chapter. On that occasion we were giving different colours (categories) of pills or no pills and we had no reason to suppose that any particular colour might produce a bigger effect than another, so a linear regression would not have been appropriate as colour in this case was arbitrary and not on any kind of linear measurement scale.

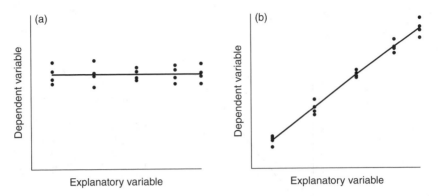

Figure 7.2   Two of possible outcomes for a linear regression for four observations of y at five levels of x. (a) The slope is equal to zero (i.e. it is parallel with the x-axis). An increase in x does not produce an increase in y. (b) The slope is not equal to zero and an increase in x produces a corresponding increase in y.

The null hypothesis for a linear regression is that the slope of the regression line does not differ from a slope with a population value of zero, that is, the line fitted to the data is parallel with the x-axis (Figure 7.2). The regression line necessarily passes through the mean of all the x values and the mean of all the y values. Classically the x-variable does not vary at random but is fixed at specific levels by the investigator and is the independent variable. We call x the *explanatory* variable (not causative!). We may use concentration as the explanatory variable for absorbance for a protein calibration curve, temperature as the explanatory variable for the speed of an enzyme reaction, or time as the explanatory variable for accumulation of an enzymatic reaction product. In these examples, we typically would select a number of concentrations, temperatures or times at which we will take our measurements, with the variable being measured as the dependent or y-variable. Because we have selected the values of x at which we will take the measurements (the values of y) we say that in this case, x is *measured without error*, meaning that it is not a random variable. Put another way, replicate measurements of y taken at a specific concentration, time, temperature, etc. have the same value of x. A practical example might be a protein calibration (standard) curve, where we decide to use six concentrations of protein standard (say 0.1, 0.2, 0.4, 0.8, 1.6 and 2.0 mg/mL) and measure the absorbance in triplicate at each concentration. At 0.1 mg/mL we will have the same value that we have assigned for x (i.e. measured without error), but the three measurements of absorbance (y) may differ, due to error such as pipetting accuracy and instrument precision, but no additional errors arising from chance variation. In this case the regression line is a line of best fit using an *ordinary least squares* (OLS) procedure and is determined mathematically by finding the fitted line that gives the minimum sum of the squared distances between the y values and the line.

In practice, for biological data, the independent variable may also be measured with error (i.e. it is a random variable) rather than fixed by the researcher, for example, body mass index of women (as the explanatory variable for infant birth weight). In this case it may be appropriate to fit the line using a slightly different method, which we will discuss a little later on.

7.2.1 Partitioning the variation

As we mentioned earlier, regression is not as simple as just taking the value of the slope and comparing it with a slope of zero because the slope with a large value may be associated with great variability in y, which may be so great that we cannot readily distinguish whether or not it differs from a slope of zero (Figure 7.3). So, we need to work the residual error in the dependent variable into the regression model. To do this, we calculate the mean of all the y values, which has the symbol $\bar{y}$, ignoring the fact that they all have different levels of x. The deviation of each y value from the mean, $\bar{y}$, is the total variation for that point (Figure 7.4, a colour version is located in the colour plate section). Some of the deviations are negative and some positive as some points lie above the regression line and some below it. If we were to sum the deviations they would equal zero, so we square them as the squares will always be positive and so the total sum of squares (SS) shall be positive. The deviation of each y value from the regression line is the variation in that data point due to residual error. Again we sum the squares of the deviations, as using the deviations themselves would equal zero, and this tells how much of the total variation in the data is due to residual error (i.e. unexplained by the regression and arising from all other random sources). The difference between the total sum of squares and the residual error sum of squares is the variation attributable to the regression.

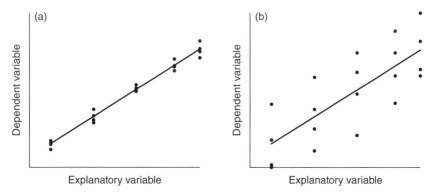

Figure 7.3 Two sets of regression data, (a) and (b), with identical slopes (2.087) but different scatter around the regression line. Slope (a) is a more accurate estimate of the value of the slope for the population than (b).

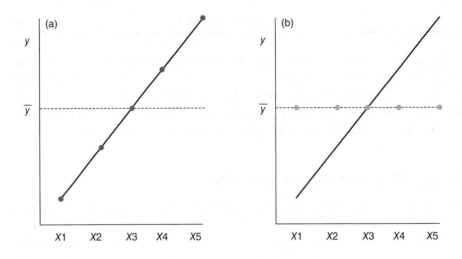

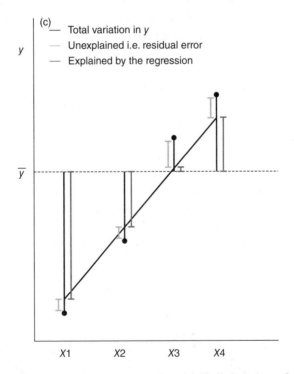

Figure 7.4 Sources of variation in linear regression. (a) If all the values of y lie perfectly on the regression line, then all the variability in y is due to the regression and there is no variation due to residual error. (b) If all the values lie on the grand mean of y, ȳ, represented by the dotted horizontal line, all the variation in y is due to residual error and there is no effect of the explanatory variable x. (c) In practice, the total variation in y comprises both regression and residual error. The total variation is represented by the blue lines and is the distance of the data points from ȳ. The variation due to residual error is represented by the distance of the data points to the black regression line (orange lines) and when subtracted from the total, gives the portion of the variation that is explained by the regression (green lines). (For a colour version, see the colour plate section.)

Where the regression sum of squares comprises a large proportion of the total variation, it is likely that there is an effect of x on y.

7.2.2 Calculating a linear regression

Now we'll calculate a linear regression using a simple data set of four points (Figure 7.5). We are going to use the ordinary least squares model (i.e. y is measured with error and x is measured without error) to demonstrate the main principles. With reference to Figure 7.4c, we need four bits of information about the data: the slope of the line, the sum of the deviations of each data point from the fitted line, how much of this total is due to the regression and how much is due to residual error.

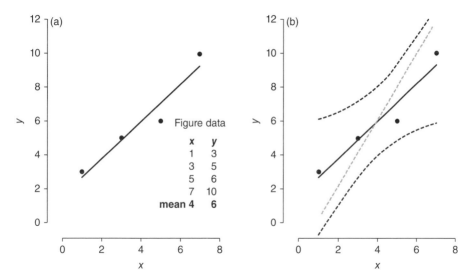

Figure 7.5 Linear regression for a simple data set of four pairs of observations. (a) The data used for the manual calculation of the linear regression in the text and (b) the 95% confidence intervals have been plotted on the graph (black dotted curved lines). The pale grey dotted line is one possible estimate of the population regression line other than the regression line (solid black) for the actual data.

The equation for the slope of the line of a least squares regression is below (Equation 7.1a). Although this looks complicated it is really quite simple. The Greek uppercase letter sigma (Σ) simply means sum or add together. I mention this because a student once asked me what the 'funny sideways M symbol means', which might be reasonable for a first-year undergraduate who had never used Microsoft Excel in 1990, but this was a medically qualified student two decades later. The symbol i means 'for each observation' and the number 1 is a reminder that each pair of observations must be kept together, x_1 with y_1 for the first pair, x_2 with y_2 for the second pair, etc. The letter n informs us that we have to sum the cross products (answers for the bracketed terms) of all the observations that we have. Calculation of the slope is a major part of the regression analysis. If we simply add a square term to the top part, the numerator, this gives us the variation in the data due to the regression, SSr (Equation 7.1b). We also need to know the amount of total variation in the data, SSt, the sum of the deviations of y from the mean of y which we also have to square as the deviations themselves will equal zero (Equation 7.1c). And finally how much of the total variation arises from random factors, the residual error, SSe. That bit is easy as we have the total and regression sum of squares, we simply subtract SSr from SSt and we have SSe.

Equation 7.1a

$$\text{slope} = \frac{\sum\limits_{i=1}^{n} (x_i - \bar{x})(y_i - \bar{y})}{\sum\limits_{i=1}^{n} (x_i - \bar{x})^2}$$

← Change in y

per

unit change in x

Equation 7.1b   Sum of squares for the regression.

$$\text{SS}r = \frac{\left(\sum\limits_{i=1}^{n} (x_i - \bar{x})(y_i - \bar{y})\right)^2}{\sum\limits_{i=1}^{n} (x_i - \bar{x})^2}$$

← Squaring the numerator of the equation above gives us SSr, or the variation in y that arises from the regression and not from random factors

Equation 7.1c   Total sum of squares.

$$\text{SS}t = \sum\limits_{i=1}^{n} (y_i - \bar{y})^2$$

The sum of squares of y is the total variation in y from the mean of y (squared so that it doesn't total 0). This is variation arising from both the regression and random factors

Now we'll put all that into practise and perform the calculations for our four data points in Figure 7.5.

Step 1. Subtract $\bar{x}$ from each value of x, square these deviations and sum them.

The mean of x, $\bar{x}$, $= 4$.

x	Deviations, dx	dx^2
1	$1 - 4 = -3$	$-3^2 = 9$
3	$3 - 4 = -1$	$-1^2 = 1$
5	$5 - 4 = 1$	$-1^2 = 1$
7	$7 - 4 = 3$	$3^2 = 9$
		SSx = 20

This is the sum of squares of x (SSx) and is the denominator (bottom half) of Equation 7.1b.

Step 2. Subtract $\bar{y}$ from each value of y, square these deviations and sum them.

The mean of y, $\bar{y}$, $= 6$.

y	Deviations, dy	dy^2
3	$3 - 6 = -3$	$-3^2 = 9$
5	$5 - 6 = -1$	$-1^2 = 1$
6	$6 - 6 = 0$	$0^2 = 0$
10	$10 - 6 = 4$	$4^2 = 16$
		SSt = 26

We have just done the calculations for Equation 7.1c. This is the total sum of squares (SSt). It is the total variation in y.

Step 3. Multiply each x deviation (from Step 1) by its corresponding y deviation (from Step 2) to give the cross products of each data point. Next sum the cross products and square the sum.

Deviations, dx	Deviations, dy	Cross products
−3	−3	9
−1	−1	1
1	0	0
3	4	12
	Sum of cross products = 22	

$$\text{Squared sum of cross products} = 22^2 = \mathbf{484}$$

The sum of the cross products is the numerator of the equation for the slope (Equation 7.1a). Its square is the numerator of Equation 7.1b. If we divide the sum of the cross products by SSx from Step 1, we will have the slope of the regression line (20/22 = +1.1). This is the slope and tells us that for each change of 1 in x, there is a 1.1 increase in y. Now we need to work out how much of the variation around the slope is due to the regression and how much is due to random factors.

Step 4.   Divide the squared sum of cross products from Step 3 by SSx from Step 1 to give the regression sum of squares (SSr).

$$SSr = 484 / 20 = \textbf{24.2}$$

This completes the calculations for Equation 7.1b.

Step 5.   Subtract the regression sum of squares from Step 4 from the total sum of squares from Step 2 to give the residual error sum of squares (SSe).

$$SSe = 26 - 24.2 = \textbf{1.8}$$

So now we have the sum of squares for the total variation, the variation due to the regression and the variation due to residual error. The final steps are the same as those for an ANOVA, we correct the SS by the degrees of freedom to give the mean squares and then calculate the F ratio.

Step 6.   Calculate the variances. As we observed for the ANOVA the regression variances are equivalent to the *mean squares* (MS). The sums of squares must be corrected for the degrees of freedom (df) to produce the mean squares. The total degrees of freedom are $n - 1$ (where n is the number of pairs of observations). The regression degrees of freedom is always 1 and therefore the error degrees of freedom is the difference, that is, total $- 1$.

The degrees of freedom for the total are the number of pairs of observations $- 1$, that is, $4 - 1 = 3$

The degrees of freedom for the linear regression is always 1

The error degrees of freedom are total df $-$ regression df, that is, $3 - 1 = 2$

To obtain the mean squares (variances) for the treatments and error we need to divide the SS from Steps 4 and 5 by the corresponding degrees of freedom.

$$\text{Regression MS} = SSr/\text{Regression df} = 24.2 / 1 = \textbf{24.2}$$
$$\text{Error MS} = SSe/\text{Error df} = 1.8 / 2 = \textbf{0.9}$$

Step 7.   Calculate the test statistic, F

$$F = \text{Regression MS/Error MS}$$
$$= 24.2 \,/\, 0.9$$
$$= 26.89$$

We can now obtain a probability value for our calculated value of F from a statistical table using the same method as the ANOVA in the previous chapter, or any statistical software will provide it for us as part of the test. The output from Minitab for these data is printed in Box 7.1. The intercept has a value of 1.6 (coefficient of the constant) and the slope of the regression line has a value of +1.10 (coefficient of x i.e. for each unit increase in x, y increases by 1.10). The standard error of the slope, 0.2121, can be used to calculate 95% confidence intervals. If we re-plot the graph for the data we can also request the software to display the 95% confidence intervals for the regression line (Figure 7.5b). The confidence interval lines allow us to visualise the range of estimated regression lines for the population (the true population regression line will lie between the two curved lines produced by 95 out of 100 samples). The grey dotted line in the figure is one possible population regression line out of many possible ones. We can see that the wider the confidence intervals the less precise the estimate of where the regression population line lies. The lines curve in to the mean value for x and y as this is the most reliable point in the data. In this case we have only four pairs of observations, so our confidence band is wide and the estimate of the population regression line is less precise. The regression is statistically significant ($p = .035$). The R^2 value is the percentage that the SSr comprises of the total SS (i.e. $24.2 \,/\, 26 * 100 = 93.1\%$). So about 93% of the variation in y is accounted for by the linear relationship (explained by x), and therefore only about 7% by residual error. It is important that the regression line should not be extended beyond the limits of the actual data as we do not know whether the relationship continues to be linear beyond the observations that we have. Extending a fitted line beyond available data is called extrapolation and is not recommended.

7.2.3   Can weight be predicted by height?

Now let's try asking a fairly simple question using some data generated by Minitab to mirror the descriptive statistics and distribution of a real data set. We might imagine that the taller someone is, the more they will weigh, simply because there is more of them in the vertical plane. Of course, we may be able to think of lots of other variables that might contribute to weight such as the width of a person's frame or the amount of fat or muscle that a person carries. So let's say that we are particularly interested in athletes, in fact swimmers

---

Box 7.1   Minitab output for regression analysis y versus x

```
The regression equation is
Response y = 1.60 + 1.10 x


Predictor          Coef       SE Coef            T          P
Constant          1.600        0.9721         1.65      0.242
X                 1.100        0.2121         5.19      0.035
S = 0.9487    R-Sq = 93.1%


Analysis of Variance

Source                 DF            SS           MS          F          P
Regression              1         24.20        24.20      26.89      0.035
Residual Error          2          1.80         0.90
Total                   3         26.00
```

---

as they train to have good general muscle strength rather than bias training to a particular part of their body (e.g. in contrast with runners who tend to concentrate on legs and javelin throwers who tend to concentrate on arms). To avoid gender differences, we decide to use the data for women. We wish to predict weight from height, so we nominate height as the independent variable (Figure 7.6). The Minitab output for the least squares regression is printed in Box 7.2.

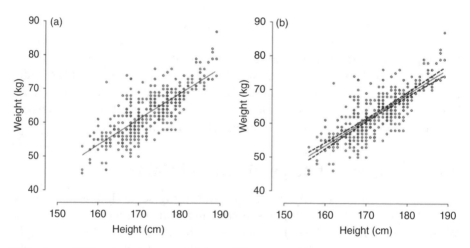

Figure 7.6   Ordinary least squares linear regression for simulated weight versus height of female swimmers $n = 360$, $F = 549.91$, $p < .001$. (a) Regression line plotted alone. (b) Regression line plotted with its 95% confidence interval (dashed lines).

Box 7.2   Minitab output for regression analysis: female swimmer weight versus female swimmer height

```
The regression equation is
Weight = - 68.3 + 0.761 Height

Predictor          Coef     SE Coef          T        P
Constant        -68.299       5.606     -12.18    0.000
Height          0.76075     0.03247      23.43    0.000

S = 4.303      R-Sq = 60.5%      R-Sq(adj) = 60.4%

Analysis of Variance

Source              DF          SS          MS          F        P
Regression           1       10164       10164     548.91    0.000
Residual Error     358        6629          19
Total              359       16793

Unusual Observations
Obs     Height     Weight        Fit     SE Fit     Residual     St Resid
  1        156     45.000     50.378      0.582       -5.378       -1.25 X
  2        156     46.000     50.378      0.582       -4.378       -1.03 X
  3        156     53.000     50.378      0.582        2.622        0.61 X
 27        162     46.000     54.943      0.410       -8.943       -2.09R
 57        166     68.000     57.986      0.310       10.014        2.33R
 75        167     74.000     58.746      0.289       15.254        3.55R
 76        167     68.000     58.746      0.289        9.254        2.16R
 92        168     70.000     59.507      0.270       10.493        2.44R
 97        162     72.000     54.943      0.410       17.057        3.98R
 98        168     73.000     59.507      0.270       13.493        3.14R
107        168     50.000     59.507      0.270       -9.507       -2.21R
110        168     73.000     59.507      0.270       13.493        3.14R
127        170     50.000     61.029      0.241      -11.029       -2.57R
170        171     72.000     61.789      0.232       10.211        2.38R
206        174     76.000     64.072      0.232       11.928        2.78R
211        174     73.000     64.072      0.232        8.928        2.08R
215        174     55.000     64.072      0.232       -9.072       -2.11R
271        177     75.000     66.354      0.270        8.646        2.01R
275        178     76.000     67.115      0.288        8.885        2.07R
287        178     58.000     67.115      0.288       -9.115       -2.12R
289        179     58.000     67.875      0.310       -9.875       -2.30R
358        189     74.000     75.483      0.581       -1.483       -0.35 X
359        189     79.000     75.483      0.581        3.517        0.82 X
360        189     87.000     75.483      0.581       11.517        2.70RX

R denotes an observation with a large standardized residual
X denotes an observation whose X value gives it large influence
```

The regression equation has a value of 0.76075 for the slope and it is statistically different from a slope of zero ($F_{1,358} = 549.91, p < .001$). The subscripted 1,358 after the F refer to the regression and error degrees of freedom, respectively, and is how the results should be reported in a publication. The intercept has a negative value, −68.299 kg. If we think about what this means in mathematical terms, it is that y (i.e. weight) is −68.299 kg when x (i.e. height) is 0 cm. This may make mathematical sense but it makes no biological sense. A female Olympic swimmer cannot have a negative weight, and neither can she have zero height. In humans, zero weight and height is not possible as even a newly fertilised ovum, the smallest possible unit a living individual once was, has both height and weight, albeit very small. If we had a different data set, we would likely have a different intercept, so the negative value we have obtained is simply a consequence of the estimate of the population regression line that this particular data set has provided and is not of biological interest. Forcing the line through the origin (0,0) in this case would be extrapolation, extending the line into values for which there is no evidence.

The linear relationship has accounted for about 60% of the variation in weight, which means that 40% of the variation remains unexplained by height. As mentioned earlier, we can probably think of some other factors that may account for the unexplained variation. Had we collected other data, such as amount of body fat and frame size, we might have been able to explain some of the remaining error using a statistical test called multiple regression, but we will move onto that subject later in this chapter.

Linear regression results summary: A linear regression revealed that height was a predictor of weight in female Olympic swimmers ($F_{1,358} = 549.91$, $p < .001$, $n = 360$). Approximately 60% of the variation in weight was explained by height (unadjusted R^2), leaving 40% of the variation unexplained. The regression equation was weight = −68.3 + 0.761 height.

Minitab also produces a list of outliers by default. Other software may or may not do this automatically. In Minitab, the section titled, Unusual Observations notifies us of any outliers, which are values with large deviations from the regression line in either axis (denoted by **R** when the large deviation is in y and by **X** when the large deviation is in the x-value). Sometimes an outlier may arise from incorrect data entry so it is worth checking the list in case this has occurred, particularly in large data sets such as this one. For example, if we found a female swimmer height of 300 cm we might be very suspicious as not even Robert Wadlow (see Chapter 4) reached that height. We could either replace the value with the correct measurement or if that is not possible, omit data for that subject. It is important that the y errors (residual errors or deviations from the regression line) are normally distributed and that they have similar variances, as these are two of the assumptions of least squares linear regression. This can be done by plotting a histogram of the standardised residuals and also a plot of standardised residuals versus the fitted values (*y* values). Recall from Chapter 5 (*z*-scores) that standardised observations have a mean of 0 and are expressed as standard deviations from the mean. Standardised residuals are the deviations of the *y* values from the regression line expressed as standard deviations. An example of how a standardised residual is calculated for the first unusual observation is given in Box 7.3 using selected Minitab output from Box 7.2.

Box 7.3   *Example of how to calculate a standardised residual using the Minitab output from Box 7.2 for a y value 45 with an x value of 156*

Note that the values displayed by the software are rounded up or down, and will make small differences to the final answer unless the same number of decimal places is used.

```
Height     Weight        Fit      SE Fit    Residual       St Resid
   156     45.000     50.378       0.582      -5.378         -1.25X
```

Using the regression equation from Box 7.2:

```
Female swimmer weight = -68.299 + 0.76075 * Female swimmer height
```

The fit is the value of weight predicted from the regression line for a height of 156 cm calculated thus:

$$\text{Fit} = -68.299 + (0.76075 * 156) = 50.378$$

The residual, or residual error, is the deviation of the observed *y* value from the value predicted from the regression line:

$$\text{Residual} = \text{observed } y - \text{fit}$$
$$= 45.000 - 50.378$$
$$= -5.378$$

The standardised residual (St Resid) is the residual divided by the square root of the residual error mean square from Box 7.2 (18.517 to three decimal places)

$$\text{St. Residual} = -5.378 / \sqrt{18.517}$$
$$= -5.378 / 4.303$$
$$= -1.25$$

For the standardised residuals to be normally distributed most of the values should occur around zero with few in the extremes (Figure 7.7a). In a plot of standardised residuals versus fitted values, the points should be distributed randomly rather than exhibit a pattern such as decreases or increases in residual error with increases in the fitted values (Figures 7.7c and 7.7d).

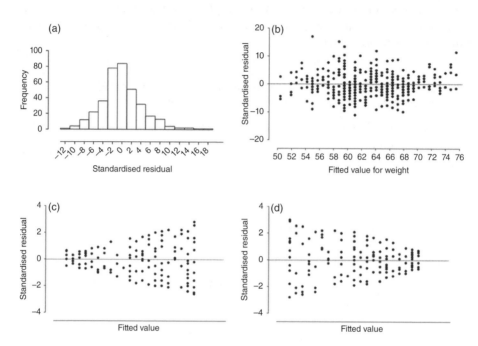

Figure 7.7 The two types of plot used to determine whether data conform to the assumptions of normality and homogeneity of variance for residual errors in a least squares linear regression. (a) Histogram of standardised residuals looks approximately normally distributed. (b)–(d) Plots of standardised residuals versus fitted values. The standardised residuals for female swimmer weight are dispersed randomly across the fitted values for weight and no pattern is discernible in (b). For (c) and (d) the data are fictitious and there is a distinct increase or decrease in variance, respectively, with increases in the fitted values, suggesting that the variances are not homogeneous.

Now I must confess that if I were to fit a line by eye to Figure 7.6a, I would place it a little steeper than the location it is given by the regression. This could be a hint that an OLS regression might not be the only way to analyse the relationship between the variables. Let's now swap the x and y variables round and plot weight on x and height on y and see if the regression line is fitted any differently through the data (Box 7.4). The value of the slope is different from that of the previous regression analysis we performed (+0.76075 for the first and +0.79560 for the second). This may seem like a small difference but its magnitude is more clearly illustrated by a graph than by numbers (Figure 7.8, a colour version is located in the colour plate section). Whether this matters or not to our analyses depends on the purpose for which we intend to use the linear regression. If our intention is to use the regression for prediction, we would obtain very different answers depending on which way round that data were plotted. We can see if this is the case by selecting fitted values to substitute into both equations. We'll use 156 cm as we have already calculated the predicted weight with the first equation and it is 50.378 kg (Box 7.3). Now let's use 50.378 kg in the second equation and if the regression line is fitted to the data in the same way, the height should be 156 cm: $122 + 0.79560 * 50.378$ $= 162$ cm, well that's 6 cm (over 2 inches) difference. The differences between the answers will be greater with values at the extremes and smaller for those nearer $\bar{y}$. This has occurred because OLS regression fits the line to minimise the vertical distances (error) in the y variable and it works just fine when x is measured without error (e.g. for a calibration curve). But in this example we have measured x with error as well. We did not select specific heights and measured the weights only of swimmers who fulfilled our height criteria; we measured both variables for all our participants.

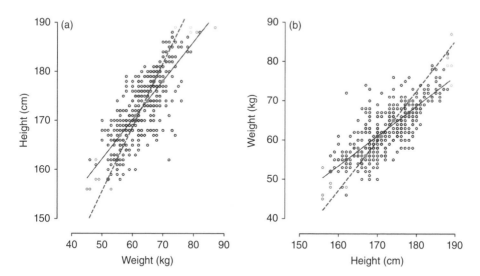

Figure 7.8 Regression for female swimmer height and weight plotted with (a) height as the dependent variable and (b) weight as the dependent variable as in the original analysis. The regression line for each plot is the solid black line. The dotted grey line is the regression line for the alternative plot, that is, in plot (a) it is the regression line for plot (b) and vice versa. To aid orientation selected data points have been coloured the same in each plot. Solid circles are those bisected by a regression line. It is clear that the two solid black regression lines do not run through the same points in each plot and hence differ in their slopes. (For a colour version, see the colour plate section.)

Box 7.4   Minitab output for regression analysis: female swimmer height versus female swimmer weight

```
The regression equation is
Height = 122 + 0.796 Weight


Predictor          Coef        SE Coef           T          P
Constant        122.437          2.150       56.95      0.000
Weight          0.79560        0.03396       23.43      0.000


S = 4.401        R-Sq = 60.5%        R-Sq(adj) = 60.4%


Analysis of Variance

Source               DF             SS          MS          F          P
Regression            1          10629       10629     548.91      0.000
Residual Error      358           6933          19
Total               359          17562
```

NOTE: UNUSUAL OBSERVATIONS HAVE BEEN OMITTED

7.2.4   Ordinary least squares versus reduced major axis regression

The problem of using an ordinary least squares regression for data where x is measured with error remains a topic of much debate (Smith, 2009). An alternative approach is to use a regression model that fits the line using the least squares of both x and y. This model is called a reduced major axis (RMA) regression and is equivalent to the geometric mean of the two slopes fitted by ordinary least squares regression (Figure 7.9). The geometric mean of two numbers is calculated by multiplying them together taking their square root (see Basic Maths for Stats section).

There is no simple answer as to which regression is most suitable for data that are measured with error in x. A decision should be based on the aims of the analysis. If the purpose is to compare the slope of the relationship between height and weight for athletes from different sports, then we need a single slope describing the relationship between these variables, which would be provided by reduced major axis. If, on the other hand, we simply wish to predict the weight of a female swimmer from their height, for example, for training purposes, then ordinary least squares would suffice keeping height as the independent variable.

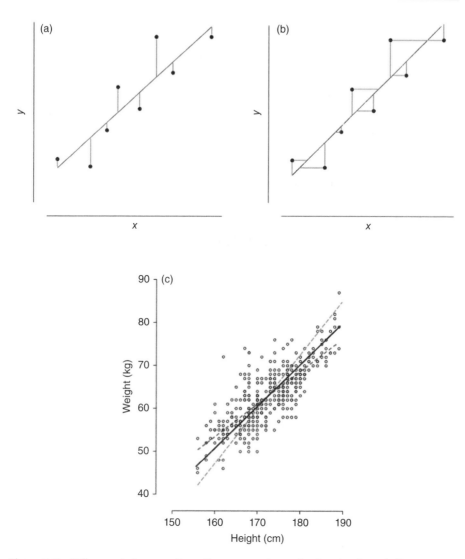

Figure 7.9 Difference between ordinary linear regression and reduced major axis linear regression. (a) For OLS the line is fitted to reduce the error in y. (b) For RMA the line is fitted to account for error in both y and x. (c) The RMA regression line (solid black) has been fitted to the female swimmer data from Figure 7.6. The dotted lines are the OLS regression lines for when height is plotted on x (blue) or weight is plotted on x (orange). (For a colour version, see the colour plate section.)

7.3 Correlation

Correlation analyses are suitable for data where y and x are each normally distributed and we wish to know the strength of an association between two continuous variables. Statisticians call these type of data a *bivariate normal distribution.* Superficially, a correlation seems similar to a regression where *x* is measure with error: both are used for paired data and plotted on a scatter-gram, and both examine how well two variables covary (proportional changes in *y* for changes in *x*), so how do they differ? In fact the same data can often be analysed using either linear regression or correlation. Students find these similarities very confusing and often do not know when to use one over the other. The answer lies in which question is being asked. If the aim of the analysis is prediction, the answer has to be linear regression because corre-lation analysis does not produce a fitted line. If the aim is simply to investi-gate whether two variables covary, a correlation can provide the answer. We'll explore this a little further with a concrete example.

7.3.1   Correlation or linear regression?

Let's use an example that we mentioned earlier to elucidate the difference between the two types of analysis. Infant birth weight is linearly related to the body mass index (BMI) of the mother before pregnancy (Koepp *et al.*, 2012). If this is all we wanted to know – whether obese mothers give birth to heavy babies and underweight mother to lighter ones – then we perform a corre-lation analysis on our data to address the question. If, however, we wish to predict the birth weight of a foetus from its mother's BMI, we would need a linear regression. Using the data from many women and the regression equa-tion, we could estimate birth weight of children born to any woman planning a family, or any pregnant woman if we have her height and pre-pregnant weight. A correlation cannot do this because there is no line of best fit and hence no regression equation from which to calculate an expected value for a variable. Furthermore, in a regression, the slope of the line informs us about the magni-tude of the effect of the *x* variable, the steeper the slope, the bigger the effect. In a correlation the 'slope' of the linear trend does not affect the strength of the correlation. We placed slope in inverted commas because a linear trend is not a fitted line but a tendency of the data to form a linear pattern. This is an important distinction between the two because the internet is teeming with 'correlation' graphs with (incorrectly) a line fitted through the data.

7.3.2   Covariance, the heart of correlation analysis

The analysis tests the strength of the linear dependence between two vari-ables. This is a measure of covariance, which is how closely a change in *x* is accompanied by a proportional change in *y* (we'll calculate it later as the deviations of *x* multiplied by the deviations of *y* corrected for degrees of free-dom). Examine the data set A in Table 7.1. Can you discern any relationship between the two variables? Multiply each value of *x* by 3 and you will find that the answer is its corresponding *y* value. If we were to plot these values on a

Table 7.1 Three sets of data with different degrees of covariance between x and y

A		B		C	
x	y	x	y	x	y
0.5	1.5	0.5	0.6	0.5	0.3
1.0	3.0	1.0	1.5	1.0	4.5
2.0	6.0	2.0	2.8	2.0	1.9
2.5	7.5	2.5	2.3	2.5	6.5
3.0	9.0	3.0	9.6	3.0	9.1
4.0	12.0	4.0	10.5	4.0	6.5

scattergram, what do you think it might look like? If you think that there will be a perfect linear relationship between x and y, you would be correct (Figure 7.10a). In this plot, x and y covary proportionally and so there is a perfect correlation between them. Now let's do the same thing for the other two sets of data in Table 7.1. The mathematical relationship between x and y is not as clear cut for these data sets, and if we plot them on a graph we can more readily visualise any trends between the two variables (Figures 7.10b and 7.10c). For data set B, there is a pretty clear linear association even though some of the points are a bit wobbly. We can discern that these data do covary, but the correlation between the variables is not as strong as that of data set A. The third data set is the least convincing of the three but there is a general trend for y to increase as x increases. For these data, the variables covary to a lesser degree and the correlation between them is weaker.

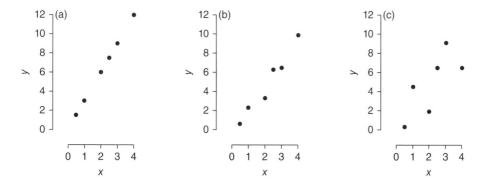

Figure 7.10 Scatter plots of the data from Table 7.1. (a) x and y covary and a linear relationship is evident. (b) x and y covary to a lesser extent but a linear relationship is still evident and (c) the covariance between x and y is less clear than in plots (a) and (b). Their tendency is towards linearity, albeit not as convincing.

7.3.3   Pearson's product–moment correlation coefficient

We can designate a number to describe the degree to which x and y covary. It is the Pearson's *product–moment correlation coefficient* but we can abbreviate it to the correlation coefficient. It is the test statistic for Pearson's correlation and is represented by the symbol r. The closer its value is to 1 or -1, the stronger the correlation (y increases more or less proportionally as x increases or decreases). The data we have just examined have the following correlation coefficients: (A) $r = 1$, (B) $r = .98$ and (C) $r = .74$, so they are all positively correlated but to different extents. The relationship between maternal BMI and infant birth weight is an example of positively correlated variables.

A negative correlation is one where y decreases as x increases. Negative correlations have a value of -1 when the two variables are perfectly negatively correlated and the closer r is to -1, the stronger the negative correlation (Figures 7.11a and 7.11b). An example of negatively correlated variables is duration of weekly exercise and percentage body fat. The more you exercise the less body fat you have. It would be nice if this were a positive correlation, but sadly losing fat requires expenditure of energy (or eating less, but that would be a positive correlation). A correlation coefficient of 0 signifies no correlation between the variables at all, and the closer r is to 0, the weaker the correlation (Figure 7.11c). When referring to the population correlation coefficient, we use the same convention as the population parameters for other parametric tests and use the Greek symbol for r, which is ρ (pronounced rho). So our sample r is an estimate of ρ. Our null hypothesis for a correlation analysis is therefore that our population correlation coefficient ρ does not differ from 0.

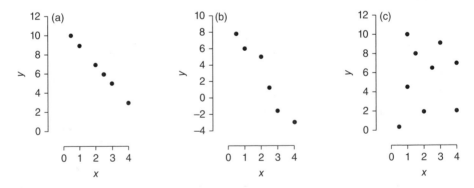

Figure 7.11 Further examples of correlations. (a) Perfectly negatively correlated variables, $r = -1$. (b) Strongly negatively correlated variables where some of the y values are negative numbers, $r = -.97$. (c) Uncorrelated variables, $r = .09$.

In summary, Pearson's correlation quantifies the strength of a linear relationship. Unlike linear regression, there is no fitted line and the variables are interdependent; there is no distinction between which is dependent and which is independent and so it does not matter which variable is plotted on x and which is plotted on y. Furthermore, Pearson's correlation cannot be used for prediction.

7.3.4 Calculating a correlation coefficient

The usual way to calculate covariance between two variables x and y is to use Equation 7.2. For an explanation of the symbols, superscripts and subscripts, please refer to the explanations provided for the linear regression.

Equation 7.2

$$\mathrm{cov}_{x,y} = \frac{\sum_{i=1}^{n} (x_i - \bar{x})(y_i - \bar{y})}{n - 1}$$

Now, the trouble with using the covariance alone is that it is scale-dependent. This simply means that if we had measurement of one variable in say, centimetres and calculated the covariance we would obtain a number that was exactly 100 times bigger than if we had expressed our measurements in metres instead of centimetres (there are 100 cm in 1 m). So the size of the numbers alters the value of covariance that we obtain. To correct for this so that measurements in one scale, for example, cm, gives the same answer as measurements in another scale, for example, m, the covariance is then divided by $n - 1$ multiplied by the standard deviation of x multiplied by the standard deviation of y. This should remind you somewhat of how we computed z-scores to standardise measurements in Chapter 5, where we divided the sum of the deviations by the sample standard deviation. In doing this we will always obtain a standardised value of r that is between -1 and 1, irrespective of how big or small our measurements are. Because the term $n - 1$ appears in both the numerator (top part) and the denominator (bottom part) and cancels out, you may be pleased to know that we can simplify the formula to Equation 7.3. Note that the numerator of the equation is the same as that for the linear regression – the key difference is in the denominator.

Equation 7.3 Pearson's correlation coefficient.

Measure of how closely a change in x is accompanied by a proportional change in y

$$r = \frac{\sum_{i=1}^{n} (x_i - \bar{x})(y_i - \bar{y})^2}{\sqrt{\sum_{i=1}^{n} (x_i - \bar{x})^2 \sum_{i=1}^{n} (y_i - \bar{y})^2}}$$

Scale-independent standardisation to a value between -1 and 1

So having tackled the previous paragraph you should be armed and ready to face calculating a correlation coefficient. We'll use a small data set to demonstrate the principle (from data set C in Table 7.1), and then we will use Minitab to perform a correlation analysis on a larger set of data for maternal BMI and infant birth weight.

Step 1. Subtract $\bar{x}$ from each value of x, square these deviations and sum them.

The mean of x, $\bar{x}$, $= 2.1667$

x	Deviations, dx	dx^2
0.5	$0.5 - 2.17 = -1.67$	$-1.67^2 = 2.79$
1.0	$1.0 - 2.17 = -1.17$	$-1.17^2 = 1.37$
2.0	$2.0 - 2.17 = -0.17$	$-0.17^2 = 0.03$
2.5	$2.5 - 2.17 = 0.33$	$0.33^2 = 0.11$
3.0	$3.0 - 2.17 = 0.83$	$0.83^2 = 0.69$
4.0	$4.0 - 2.17 = 1.83$	$1.83^2 = 3.35$
		SSx = 8.34

We'll use the deviations, dx, in both numerator and denominator later and the sum of squares of x, SSx in the denominator as part of the calculation for the standard deviation of x. SSx is the same as $\sum (x - \bar{x})^2$ in Equation 7.2.

Step 2. Subtract $\bar{y}$ from each value of y, square these deviations and sum them.

The mean of y, $\bar{y}$, $= 4.80$.

y	Deviations, dy	dy^2
0.3	$0.3 - 4.80 = -4.50$	$-4.50^2 = 20.25$
4.5	$4.5 - 4.80 = -0.30$	$-0.30^2 = 0.09$
1.9	$1.9 - 4.80 = -2.90$	$-2.90^2 = 8.41$
6.5	$6.5 - 4.80 = 1.70$	$1.70^2 = 2.89$
9.1	$9.1 - 4.80 = 4.30$	$4.30^2 = 18.49$
6.5	$6.5 - 4.80 = 1.70$	$1.70^2 = 2.89$
		SSy = 53.02

We'll use the deviations, dy, in both numerator and denominator later and the sum of squares, SSy, in the denominator as part of the calculation for the standard deviation of y. SSy is the same as $\sum (y - \bar{y})^2$ in Equation 7.2.

Step 3. Calculate the cross products of the deviations dx and dy.

Deviations, dx		Deviations, dy		Cross products
−1.67	*	−4.50	=	7.52
−1.17	*	−0.30	=	0.35
−0.17	*	−2.90	=	0.49
0.33	*	1.70	=	0.56
0.83	*	4.30	=	3.57
1.83	*	1.70	=	3.11

Sum of cross products = 15.60

The sum of the cross products is the numerator of the equation.

Step 4. Calculate the standard deviation of x * standard deviation of y using SSx from Step 1 and SSy from Step 2.

$$= \sqrt{8.34 * 53.02}$$
$$= \sqrt{442.1868}$$
$$= 21.03$$

Step 5. Divide the answer for the numerator from Step 3 by the answer for the denominator from Step 4.

$$r = 15.60 / 21.03$$
$$= 0.74$$

Congratulations, this is the correlation coefficient.

7.3.5 Interpreting the results

Now that we have a correlation coefficient we can do two things: we can square it to give r^2 and we can derive a probability value. The probability can be derived from a table of critical values for Pearson's r (Table 7.2). The degrees of freedom for Pearson's correlation is $n - 2$, where n is the number of pairs of observations, we have six, so we have four degrees of freedom. If we check in the table, for $\alpha = .05$, our calculated value of r must exceed 0.811, which it does not, so the correlation for these data is not statistically significant and we fail to reject the null hypothesis that r does not differ from zero. The square of the correlation coefficient informs us of how much of the variation in x and y results from the linear relationship between them, and how much from random factors. For this set of data, $R^2 = 0.74^2 = 0.55$. We can multiply this by 100 to express it as a percentage, so 55% of the scatter in the data, just over half, arises from the correlation but quite a large part, 45%, is attributable to other sources. A correlation coefficient of 0.74 may appear to be a fairly strong correlation, but where the deviations from linearity are large and the sample size is small, as in this case, the scatter in the data does not provide a good enough estimate of the population correlation coefficient. Had we had just two more observations with the same value of r, we would just have had a large enough sample size to provide a better estimate as critical r would have been .707 (Table 7.2).

Table 7.2 Critical values for Pearson's r for up to 6 degrees of freedom

df	$\alpha = .05$	$\alpha = .01$
1	0.997	0.999
2	0.950	0.990
3	0.898	0.959
4	0.811	0.917
5	0.754	0.874
6	0.707	0.834
...	...	...

7.3.6 Correlation between maternal BMI and infant birth weight

Now let's use a larger data set to examine whether maternal BMI and infant birth weight are correlated. The data have been randomly generated to mirror the descriptive statistics published in Koepp *et al.* (2012). These authors studied a huge number of women (58,383) but we are going to use a much smaller sample size of 100 (Figure 7.12). The Minitab output for these data is printed in Box 7.5. You may be relieved that this output is far less complicated than that for any of the tests we have encountered thus far and requires a little explaining. The correlation coefficient is 0.825 which is statistically significant, $p < .001$. If we calculate $R^2 = 0.825^2 = 0.681$, so 68.1%, or just over two thirds of the variation is due to the linear relationship and about a third from random factors.

Box 7.5   *Minitab output for correlations of BMI, birth weight*

Pearson correlation of BMI and birth weight is 0.825

p-Value = 0.000

RECALL THAT MINITAB DOES NOT REPORT PROBABILITIES TO MORE THAN THREE
DECIMAL PLACES SO THIS SHOULD BE REPORTED AS $p < .001$

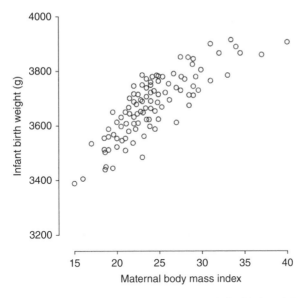

Figure 7.12 Scattergram of maternal pre-pregnant BMI and the birth weight of the newborn infant, *n* = 100. As the data are being analysed with a Pearson's correlation, the data could have been plotted the other way round with infant birth weight on *x*.

7.3.7 What does this correlation tell us and what does it not?

BMI is a measure of obesity: 25–29 is considered overweight and 30 or above is classified as obese. Our correlation analysis above informs us that there is a fairly strong correlation between a woman's pre-pregnant BMI and the weight of her newborn infant. What it does not mean is that a high BMI *causes* bigger babies. A higher maternal BMI may well play a causal role in increased infant birth weight *but the correlation cannot test this.* Furthermore, the causes may be many and they may be quite different. Think of it this way, there is a very strong correlation between a child's height and their ability to do maths, but nobody in their right mind would suggest that height causes a child to have a greater mathematical ability. Other factors such as age, vocabulary, brain development, education and experience that are more likely to contribute to their ability to do maths increase with a child's height.

Pearson's correlation does not allow us to predict the weight we might expect the baby to be at birth from the mother's BMI. If we wanted to do this, we would need to use a linear regression. It also does not tell us how big the effect of BMI is on infant birth weight, again for this we would need a linear regression and this information would be provided by the slope of the regression line. In the original paper by Koepp *et al.* (2012), the data from

58,383 women were in fact examined using linear regression, which confirms our earlier affirmation that data can often be analysed using either method. The key to selecting which test to use is in clearly defining the question that you wish to answer.

Pearson's correlation results summary: A strong positive correlation was observed between maternal BMI and infant birth weight ($r = .85, p < .001$, $n = 100$). Approximately 68% of the variation was explained by the correlation (unadjusted R^2).

7.3.8   Pitfalls of Pearson's correlation

Once its principles are understood, correlation is not a complicated statistical test to interpret as there are only two variables and no treatment groups, unlike ANOVA or multiple regression. However, Pearson's correlation suffers from a number of potential problems. Where the sample size is very large, the value r can be very small and yet be statistically significant. However, statistical significance does not necessarily equate to data being biologically meaningful. A statistically significant correlation coefficient of 0.4 with a large sample size means that only 16% ($0.4^2 * 100 = 16$) of the variation is due to the correlation so most is due to other factors or random error. It is unlikely that this correlation would be of any interest biologically, despite its statistical significance.

Another problem with Pearson's correlation is that it is very sensitive to the effects of outliers, data points with large deviations from the means of x and/or y relative to the rest of the data and which carry undue influence. Just a single outlier can produce a statistically significant correlation where none exists without it (Figure 7.13). The data in Figure 7.13a are exactly the same as those plotted in Figure 7.11c just with different maxima on the axes and are not correlated – the value of r is only 0.088. Adding just one point produces an allegedly significant correlation between the two variables. It's clear here that there is a problem but what can we do if we have many data points with a few possible 'outliers'? Well, as we mentioned earlier each variable for Pearson's correlation should conform to a normal distribution so we can test for normality of each variable (see Chapter 6, under assumptions of the t-test). If points deviate markedly from the line in an Anderson–Darling probability plot, then we have outliers. It is not recommended removing the outlying points, but there is a non-parametric test that we could use instead – Spearman's rank correlation test (see also Chapter 9). By reducing the absolute values to ranks, the effect of data with extreme values is negated.

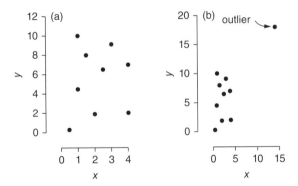

Figure 7.13 The effect of a single outlier on Pearson's correlation coefficient. (a) Uncorrelated data replicated from Figure 7.11c but with axes scaled to facilitate comparison with the outlier in b, $r = .088$. $p = .821$. (b) The same data with a single outlier added. There now appears to be a statistically significant correlation $r = .751$, $p = .012$.

The data for Figure 7.13b have been ranked lowest to highest in Table 7.3 and the ranked and original data plotted on scattergrams (Figure 7.14). Note that on the plot of ranked data the outlier (open circle) is no longer such but is much closer to the other data points, so the influence it has is much reduced. The pattern of the other data points has not been greatly affected by ranking them as they were not correlated originally (compare Figure 7.14a with Figure 7.14c). The Spearman's test is identical to Pearson's, except that it is performed on the ranked values rather than the original values. A little confusingly it produces a test statistic that is also called r, so it should be reported as Spearman's r or Spearman's correlation coefficient, rather than Pearson's r.

Table 7.3 Original and ranked values of x and y for the data from Figure 7.14b

x	x ranked	y	y ranked
0.5	1	0.3	1
0.9	2	4.5	4
1.0	3	10.0	9
1.5	4	8.0	7
2.0	5	1.9	2
2.5	6	6.5	5
3.0	7	9.1	8
3.9	8	7.0	6
4.0	9	2.0	3
14.0	10	18.0	10

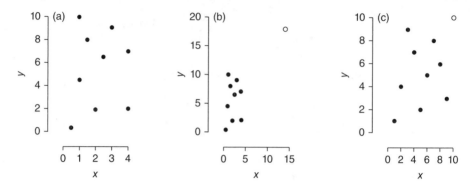

Figure 7.14 Spearman's rank correlation reduces the influence of outliers by using ranked data. (a) Uncorrelated original data plotted without the outlier, $r = .088$. $p = .821$. (b) The same data with the single outlier (open circle) added, $r = .751$, $p = .012$; the single outlier produces a statistically significant correlation due to its extreme values. (c) The ranked data with outlier (open circle) present $r = .394$. $p = .260$. The influence of the outlier is now much reduced permitting the correct conclusion to be made that the two variables are not correlated.

7.4 Multiple regression

Most economists think of God as working great multiple regressions in the sky.

Edgar R. Fiedler

We have learned that correlation is a useful statistical technique widely used to assess the linear relationship between two variables. The correlation coefficient tells us the extent to which points in a scatter plot conform to a straight line, and by its polarity whether the relationship is positive or negative. The closely related technique of regression quantifies the relationship between the variables by providing an equation for the linear relationship in terms of slope and intercept, and therefore allows the prediction of values. Multiple regression takes the analysis one step further by allowing more than one independent variable to be used to predict (or explain) the dependent variable on *y*. Using multiple predictors reflects more accurately the true relationship between real variables – usually the phenomenon in which we are interested will have more than one explanatory variable. Multiple regression is a flexible statistical technique with wide general applicability. However, perhaps because of this flexibility, it is open to misuse and abuse.

We have made up a small data set to illustrate how the technique can be useful (Table 7.4). Let us say we recruited some athletes, mainly elite sporty types (elite), but to make up numbers, a few others whose only interest in sport is watching it on TV (couch potatoes). We are interested in heart rate changes (dependent variable) in response to the type of activity and intensity of exercise (independent variables). Each participant was randomly selected to carry out a specified type of activity (walk, jog, sprint), while intensity was

Table 7.4 Raw data for measurements of heart rate at different levels of exercise intensity in 18 study participants

Heart rate	Intensity	Activity	Athlete
30	7	Walk	Elite
35	8	Walk	Elite
50	9	Walk	Elite
32	8	Walk	Elite
45	10	Jog	Elite
50	11	Jog	Elite
55	12	Jog	Elite
52	11	Jog	Elite
58	10	Jog	Elite
65	11	Sprint	Elite
70	12	Sprint	Elite
75	14	Sprint	Elite
70	4	Walk	Couch potato
75	6	Walk	Couch potato
85	7	Jog	Couch potato
110	9	Sprint	Couch potato
95	9	Sprint	Couch potato
100	8	Sprint	Couch potato

measured by the exercise machine on a scale from 0 to 20. Heart rate was measured at the end of the specified activity.

We have first performed a simple linear regression on these data and plotted them on a graph (Figure 7.15). The linear regression line for these data has a negative slope, −0.41, and the regression is not statistically significant ($F = 0.029$, $p = .867$). A Pearson's correlation between heart rate and exercise intensity gives $r = −.043$, $p = .867$. That's right, like the regression line, the relationship is slightly negative. If we examine Figure 7.15 more closely, there appear to be increases in heart rate as exercise intensity increases so why do we have a negative relationship when it looks positive by eye? Well, note that there are two clusters of points – and these clusters correspond to the two types of athlete (couch potatoes – upper left, elite – lower right). We have here a problem known as *heterogeneity of subsamples* (groups within the data), and this invalidates the simple analysis. However, if we include the type of athlete, and for good measure the specified activity, into a multiple regression analysis, then we can examine the effects of more than one independent variable on the dependent variable heart rate. What we see in the plot is that within each cluster there is a clear positive relationship between heart rate and exercise intensity. However, elite athletes have lower heart rates and can work the exercise machine harder. So the apparent negative relationship between heart rate and exercise intensity is due to including two different (heterogeneous) types of athlete in our sample. The R^2 statistic measures how much

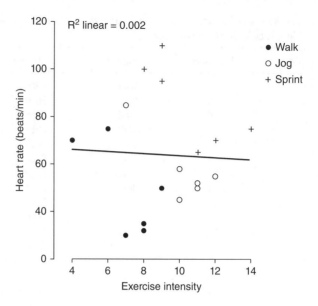

Figure 7.15   Ordinary least squares linear regression with exercise intensity and the independent variable and heart rate as the dependent variable in a fictitious study of elite athletes and couch potatoes. The regression equation is heart rate = 67.7 − 0.41 ∗ exercise intensity.

variability is explained by the relationship between the variables, and here it is negligible at 0.2% (Figure 7.15). We need to improve our analysis ... this is where a multiple regression is useful.

There are three general ways of performing a multiple regression: standard, sequential (hierarchical) and statistical (stepwise). For detailed guidance on which to use you should consult Tabachnick and Fidell (2007), which is an excellent advanced reference for multiple regression and other multivariate methods. Here, for simplicity, we will use standard multiple regression, entering all three independent variables (exercise intensity, activity and type of athlete) into the analysis simultaneously. All statistical packages will do the analysis but we will use SPSS to illustrate. The equivalent Minitab and R outputs are located in Appendix B. The SPSS output, with various options selected, is reported in Box 7.6.

SPSS provides a lot of information, but we are going to concentrate only on the essentials, as you should, having read the earlier part of this chapter, you know how to interpret some of the other information such as sums of squares, df, etc. The overall relationship, between the three independent variables and the dependent variable, is now statistically significant (Box 7.6, ANOVA section) with $F_{3,14} = 103.3, p < .001$. The numbers given after F here represent the degrees of freedom for the regression with 3 predictors and 14 df for the residual ($n - 1 - 3$). R^2 has dramatically improved from 0.2% to 95.7% (Box 7.6, Model summary), hence most of the variation in the data is attributable to the

Box 7.6 *SPSS output for a standard multiple regression performed on the data in Table 7.4. For the independent variable type of activity, the data were coded as 1 – walk, 2 – jog, 3 – sprint, and for independent variable type of athlete they were coded as 0 – elite, 1 – couch potato*

Model Summary[b]

Model	R	R Square	Adjusted R Square	Std. Error of the Estimate
1	.978[a]	.957	.947	5.320

a. Predictors: (Constant), Type of athlete, Activity, Exercise intensity
b. Dependent Variable: Heart rate

ANOVA[b]

Model		Sum of Squares	df	Mean Square	F	Sig.
1	Regression	8767.734	3	2922.578	103.254	.000[a]
	Residual	396.266	14	28.305		
	Total	9164.000	17			

a. Predictors: (Constant), Type of athlete, Activity, Exercise intensity
b. Dependent Variable: Heart rate

Coefficients[a]

Model		Unstandardized Coefficients		Standardized Coefficients	t	Sig.	95.0% Confidence Interval for B		Collinearity Statistics
		B	Std. Error	Beta			Lower Bound	Upper Bound	VIF
1	(Constant)	.519	9.677		.054	.958	-20.236	21.274	
	Exercise intensity	3.271	1.436	.344	2.278	.039	.191	6.352	7.369
	Activity	9.061	3.328	.328	2.723	.017	1.923	16.199	4.695
	Type of athlete	45.572	5.822	.952	7.827	.000	33.084	58.059	4.790

a. Dependent Variable: Heart rate

regression. In the coefficients section, we can see that each of the independent variables is now statistically significant $p < .05$, and reassuringly, there is now a positive relationship between heart rate and exercise intensity (unstandardised slope coefficient, B). The slope coefficient for the type of athlete is large and positive, and represents the large 'step' difference between elite athletes and couch potatoes (coded 0 and 1, respectively, in the analysis). Just as with simple regression we write out the equation, here it would be:

$$\text{Heart rate} = 0.519 + 3.271\,(\text{Exercise intensity}) + 9.061(\text{Activity})$$
$$+\, 45.572\,(\text{Type of athlete})$$

What multiple regression does is to calculate the optimal relationship between a combination of predictor independent variables and the dependent variable (by minimising the squared residuals). Each variable's coefficient in the equation tells us how much the dependent variable changes for each unit (1) increase of that independent variable, while keeping all other independent variables constant. Saying an independent variable is a 'predictor' does not mean that it has a causal relationship with the dependent variable.

Regression analyses are characterised by numerous statistical diagnostic tests. Initially these appear to be tedious formalities, however, with time and experience their usefulness and importance is increasingly valued. One such diagnostic is the *collinearity* (also known as *multicollinearity*) statistic given in the last column of the coefficients section of Box 7.6. VIF stands for variance inflation factor, and informs us that the standard errors for each of the variable coefficients are inflated by between 4.695 and 7.369 times. An alternative statistic usually given is the *tolerance* which is merely the reciprocal of the VIF (i.e. 1/VIF). Inflation of standard errors indicates instability of the regression equation, and arises because two or more independent variables are strongly correlated (either positively or negatively) with each other (Box 7.7). A rule of thumb is that you should be very concerned if there is a VIF > 10 (or tolerance <0.1), and act to remove one or more independent variables. In this case we should be a bit concerned about the 7.369 associated with exercise intensity. Since we are particularly interested in this variable as a predictor of heart rate, we should consider removing another independent variable rather than this one. Clearly, type of athlete is also crucial (that's how we got into doing the multiple regression to avoid heterogeneous subsamples), so we could consider removing the type of activity. It is quite strongly correlated with exercise intensity ($r = .603$, Box 7.7), moreover, its standardised coefficient at 0.328 is the smallest – which means it is the weakest among the three predictors. Although it is often easy to find statistical reasons to include or remove independent variables, the best reasons come from your understanding of the importance of particular variables to the analysis.

> **Box 7.7   Edited SPSS output of Pearson's correlation coefficients for all combinations of variable pairs using the data in Table 7.4. For activity and type of athlete the data were coded as in Box 7.6**
>
> Correlations
>
> | | | Exercise intensity | Activity | Type of athlete |
> |---|---|---|---|---|
> | Heart rate | Pearson Correlation | −.043 | .672 | .789 |
> | | p value (2-tailed) | .867 | .002 | .000 |
> | Exercise intensity | Pearson Correlation | | .603 | −.613 |
> | | p value (2-tailed) | | .008 | .007 |
> | Activity | Pearson Correlation | | | .144 |
> | | p value (2-tailed) | | | .568 |

Variables that are of theoretical importance, even if not statistically significant, should still be included in a multiple regression. Let us say here, for illustrative purposes, that we are primarily interested in exercise intensity and so remove the variable 'type of activity'. The SPSS output for this analysis is given in Box 7.8.

We can see that R^2 is still very large at 93.4% (Box 7.8, Model Summary section). The VIF are now much lower, giving narrower 95% confidence intervals for the variable B values (Coefficients section) compared with those of the 'all in' analysis shown in Box 7.6. Furthermore, the standardised coefficients (beta) are higher and more statistically significant (both $p < .001$).

Box 7.8   SPSS output of second regression analysis dropping type of activity from the analysis. Details otherwise as in Box 7.6

Model Summary

Model	R	R Square	Adjusted R Square	Std. Error of the Estimate
1	.966^a	.934	.925	6.357

a. Predictors: (Constant), Type of athlete, Exercise intensity

ANOVA^a

Model		Sum of Squares	df	Mean Square	F	Sig.
1	Regression	8557.915	2	4278.957	105.900	.000^b
	Residual	606.085	15	40.406		
	Total	9164.000	17			

a. Dependent Variable: Heart rate
b. Predictors: (Constant), Type of athlete, Exercise intensity

Coefficients^a

Model		Unstandardized Coefficients B	Std. Error	Standardized Coefficients Beta	t	Sig.	95.0% Confidence Interval for B Lower Bound	Upper Bound	Collinearity Statistics Tolerance	VIF
1	(Constant)	-17.571	8.406		-2.090	.054	-35.488	.346		
	Exercise intensity	6.731	.800	.707	8.410	.000	5.025	8.436	.624	1.603
	Type of athlete	58.502	4.024	1.222	14.539	.000	49.926	67.079	.624	1.603

a. Dependent Variable: Heart rate

Our regression equation is now:

Heart rate $= -17.571 + 6.731$ (Exercise intensity) $+ 58.502$ (Type of athlete)

We could predict the heart rate of a couch potato (coded 1) doing moderate exercise (say intensity of 9) by substituting these values into the regression equation:

Heart rate $= -17.571 + 6.731(9) + 58.502(1) = 101.510$ beats per minute

Multiple regression results summary: In a sample of 18 participants (12 elite athletes and 6 couch potatoes) exercise intensity, type of activity (walk, jog, sprint) and the type of athlete were studied as predictors of heart rate. All three predictor variables were significantly related to heart rate, however, there was moderately high collinearity between the variables. Box 7.7 shows the correlations between the variables. Our primary interest was exercise intensity and type of athlete, so only these were included in a re-run of the regression. There was a statistical relationship between the predictor variables and heart rate, $F_{2,15} = 105.90, p < .001$, and accounted for 93.4% of the variability (unadjusted R^2). The equation for the relationship was

Heart rate $= -17.571 + 6.731$ (Exercise intensity)
$+ 58.502$ (Type of athlete)

Both predictor variables contributed significantly $p < .001$: exercise intensity beta $= 0.707$ and type of athlete beta $= 1.222$. A three-dimensional (3D) plot of the variables is given in Figure 7.16.

With one independent variable, the relationship with the dependent variable is (2D) about a fitted line, with two independent variables we can graph the fit in three dimensions with data points scattered about a plane. In our example, two clusters of points occur at different places about a plane, each cluster determined by the type of athlete (Figure 7.16). With three or more independent variables the points are scattered in hyperspace.

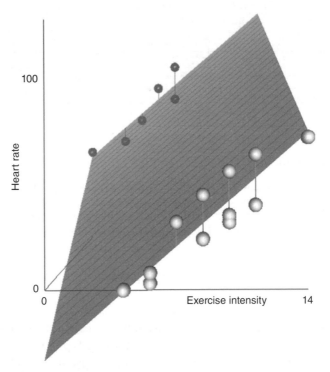

Figure 7.16   Three-dimensional plot of the dependent variable heart rate and two independent variables, exercise intensity on *x* and type of athlete on *z*. The type of athletes is also indicated by the different shades and sizes (distance) of the points: dark and small are the couch potatoes, and light and large are the elite athletes.

In order to visualise the relationship of an individual independent variable and dependent variable it is possible to construct a *partial regression plot*. Of particular interest to us is intensity of exercise as a predictor of heart rate (Figure 7.17). This partial regression plot represents the relationship between heart rate and exercise, simultaneously taking into account the type of athlete. Here we see a cigar-shaped and clear positive relationship between these variables (as we had expected).

Multiple regression is a useful general technique for analysing data where the dependent variable (also known as an *outcome* or *criterion* variable) is associated with more than one independent variable (also known as an

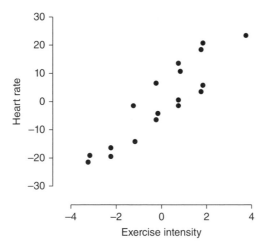

Figure 7.17 Partial regression plot. If the effect of different types of athlete is kept constant, then we can see that there is a clear positive linear relationship between heart rate and exercise intensity here. Scales are centred on the means. Such plots can be produced in SPSS during regression analysis.

explanatory or *predictor* variable). It can handle continuous and dichotomous independent variables (and indeed the dependent variable can also be dichotomous, as in *logistic regression*). Regression can be performed instead of ANOVA (by dummy variable coding the different levels of a factor), but ANOVA cannot necessarily be performed using regression data if one or more variables are continuous (though it could be done by converting to categories e.g. low/med/high – but with loss of information). ANOVA is a restricted form of regression. Actually our particular example could have been analysed in a between-participants ANOVA, entering type of athlete as a fixed factor and exercise intensity as a covariate, but it would not normally have produced the coefficients used to construct the regression equation.

When considering data for analysis by multiple regression, you should carefully select which variables to enter, rather than just enter them all. Sometimes attempts are made to find the best equation by entering as many independent variables as possible, regardless of their meaning. Unfortunately, this is encouraged by stepwise procedures (see Tabachnick and Fidell (2007)), and should be avoided, except perhaps for exploratory analyses. It is important to stress that for an accurate regression equation, all relevant independent variables with respect to dependent variable changes must be included in an analysis. Imagine if we had not included the type of athlete variable in our example (and had not recorded it), we would have been unable to interpret our data. If there are more independent variables than cases, then the regression equation predicts precisely the dependent variable values. We could have included sex, type of sport, age, height, weight, health status, etc. as predictors. This would lead to a better fit to the particular sample data, but paradoxically leads to a less useful result because of *overfitting* to the idiosyncrasies of our particular sample. This means that the results of our analysis might not be readily applicable to other data selected from the same population, that is, there is poor generalisation. One way to check how 'good' the regression equation is, is to apply *cross-validation* to the data. Here, a regression equation developed from a randomly selected large subset of the data is then used to predict scores from the remaining data. The predicted and actual scores are correlated, and the R^2 compared with the initial R^2 from the larger subset. We would expect the former value to be similar though slightly lower than the latter.

In some situations it may be appropriate to investigate an interaction between independent variables. Recall from the last chapter where there could be an interaction in the two-way ANOVA. The same can happen in multiple regression. In the ANOVA the interaction is automatically computed by the default analysis, whereas this is not done automatically in multiple regression. You will need to create another variable that represents the interaction. This is done quite simply, by multiplying two (or more for three-way or higher interactions) of them together, giving you another column of data with the products. This new interaction variable is then entered into the analysis along with the individual independent variables. Including interaction terms in your analysis can produce multicollinearity. This may be avoided by *centring* them. Centring is done by converting each value to a deviation from the mean, so that their mean value is zero. It is not necessary to centre the dependent variable. If an interaction appears to be important then it is helpful to do a 3D plot of the individual variables comprising the interaction in order to visualise the relationship and understand the interaction. If one of the predictor variables consists of a few levels, then a 2D plot can be done with different lines representing the different levels for that variable. As you can imagine, it becomes more complicated visualising an interaction between three variables – especially when all variables consist of continuous measurement data (see Aiken and West, 1991 for details). In practice, most studies would only consider two-way interactions at most.

As emphasised earlier in this chapter (Figure 7.7), diagnostic plots of residuals are crucial. If the residuals are not normally distributed (Figure 7.7a), then that might indicate that one or more variables (independent and dependent) need to be transformed. If the residuals are not randomly scattered (Figure 7.7b), then that might indicate that another variable is in play (e.g. the order in which the data were collected), or another variable you just did not think of (e.g. gender). Including this variable in another analysis should then result in improved diagnostic plots. An important assumption is *linearity*. This is particularly important when categorical variables are included in the analysis. If a predictor with m levels is non-linear with respect to the dependent variable, it can be converted into $m - 1$ *dummy* variables (coding each with 0 and 1), ensuring linearity since a straight line always connects between the two points $0,y_1$ and $1,y_2$. As an example, we could do this for type of **activity** variable which has three levels (perhaps the relationship between these levels and heart rate is not linear, e.g. the heart rates produced by sprinting are very much higher than for the other two activities). We will use one of the levels as our *base* condition. Let us say for convenience this is the **walk** condition. This means that all the other conditions (**jog** and **sprint**) will be compared with it. The two additional dummy variable columns are shown in Table 7.5. The **walk** condition does not need a variable column as it is the base condition. The two new columns then code for the other two activities. Hence, the first four rows refer to **walk** and therefore both dummy columns contain '0's. Rows 5–9 code for **jog** and therefore contain '1's in the **jog** dummy column and '0's in the **sprint** dummy column. The next three rows contain '1's in the **sprint** column and '0's in the **jog** column, etc. The regression analysis can be done including these two new variables in place of the type of **activity** variable.

Table 7.5 This is the same as Table 7.4 with the addition of two additional columns that represent the dummy coding for activity

				Dummy variables	
Heart rate	Intensity	Activity	Athlete	Jog	Sprint
30	7	Walk	Elite	0	0
35	8	Walk	Elite	0	0
50	9	Walk	Elite	0	0
32	8	Walk	Elite	0	0
45	10	Jog	Elite	1	0
50	11	Jog	Elite	1	0
55	12	Jog	Elite	1	0
52	11	Jog	Elite	1	0
58	10	Jog	Elite	1	0
65	11	Sprint	Elite	0	1
70	12	Sprint	Elite	0	1
75	14	Sprint	Elite	0	1
70	4	Walk	Couch potato	0	0
75	6	Walk	Couch potato	0	0
85	7	Jog	Couch potato	1	0
110	9	Sprint	Couch potato	0	1
95	9	Sprint	Couch potato	0	1
100	8	Sprint	Couch potato	0	1

A number of diagnostics help detect outliers which might exert excessive influence (known as *leverage*) within the equation. It may be that these outliers are errors, in which case they should be corrected, if possible, or removed. If there are specific justified reasons for excluding them, then that can be done judiciously, though analyses done with and without them included should be compared. If the outliers are genuine and problematic, then you should consider doing a transform on the variable.

Finally, it is necessary to say something about sample size. Our fabricated example was clearly deficient. The *effective* sample size will be reduced if there is multicollinearity present among the predictor variables. Generally, with medium-sized effects, you will need at least 50 participants + (8 times the number of independent variables). So in our example, we would need $50 + (3 * 8) = 74$. Otherwise, the more the better.

7.5   Summary

- Paired data that conform to a bivariate normal distribution, may often be analysed using either linear regression or correlation analyses. Both are tests of a relationship between the x and y variables, but the choice of which to use depends on the hypothesis one wishes to test.

- Plotting data on a scattergram is a useful first step in deciding whether testing for a linear association is reasonable. If data are clearly curvilinear or have no obvious tendency towards linearity, neither test is appropriate.

- For an OLS regression, x is classically measured without error, but linear regression is commonly performed when both variables are measured with error. A judgement must therefore be made as to whether an OLS or an RMA line fit is the more informative depending on other considerations, whether prediction is anticipated or comparison with similar data sets may be desired.

- Pearson's correlation does not permit prediction as there is no fitted line, although it is a measure of the strength of the linear relationship between two variables.

- Multiple regression offers the ability to incorporate multiple predictor variables and control for confounds. Often variables are multiply linked to many other variables, so this is a more realistic way of looking at the relationship between particular variables and their outcome.

References

Aiken, L.S. and West, S.G. (1991) *Multiple Regression: Testing and Interpreting Interactions*. Newbury Park, CA: Sage Publications.

Koepp, U.M.S., Andersen, L.F., Dahl-Joergensen, K., Stigum, H., Nass, O. and Nystad, W. (2012) Maternal pre-pregnant body mass index, maternal weight change and offspring birthweight. *Acta Obstetrica Gynecologica Scandinavica*, 91(2):243–249.

Smith, RJ. (2009) Use and misuse of the reduced major axis for line fitting. *American Journal of Physical Anthropology*, 140(3):476–486.

Tabachnick, B.G. and Fidell, L.S. (2007) *Using Multivariate Statistics*, Boston, MA: Pearson Education.

8

Analysis of Categorical Data – 'If the Shoe Fits ...'

8.1 Aims

Much of the data which we are familiar with and which is collected in research consists of things (people, animals, observations) which are consigned to specific categories. Each category has so many observations. There are a number of different statistical tests that we can do to see whether the number of observations in each of the categories is what we would expect according to our hypothesis, typically a null hypothesis. The simplest scenario consists of just one variable containing two or more different categories (levels). Analysis of these data is concerned with how well the data fit according to the hypothesis. More complex is a scenario consisting of two variables where each variable has two or more different levels. Here, we are interested in whether there is an association between the two variables. According to the null hypothesis, the two variables are independent of each other, so we would be interested in the relative proportions across the two variables. Epidemiology consists of numbers of people improving or worsening in therapy or disease. The data in such studies are also represented by categorical data.

8.2 One-way chi-squared

Data consisting of frequencies of events assigned to different classes are known as categorical or nominal data. In these circumstances we are interested in the relative proportions in the different categories. Using a questionnaire it is possible to classify people according to their handedness. Most

Starting Out in Statistics: An Introduction for Students of Human Health, Disease, and Psychology
First Edition. Patricia de Winter and Peter M. B. Cahusac.
©2014 John Wiley & Sons, Ltd. Published 2014 by John Wiley & Sons, Ltd.
Companion Website: www.wiley.com/go/deWinter/startingstatistics

people are right-handed, that is, their right hand dominates in skilled performances like writing, cutting or playing a racket sport. A few people use their left hand for everything, and mixed handed people use both left and right hands to different extents. A study by DeLisi *et al.* (2002) looked at handedness in patients with schizophrenia. A total of 418 participants were categorised into three groups: right handed (348), mixed handed (42) and left handed (28). If there were no particular hand preference across the participants we would expect equal numbers in each group. Clearly, that's not what they found. Like those unaffected by schizophrenia, there were far more right handers and fewest left handers. To do a statistical analysis we need to specify a null hypothesis that participants are equally likely to be in any of the three categories. If the null hypothesis were true then we would expect one third of our sample to be right-handed, a third mixed handed and a third left-handed, hence there would be an expected frequency of 139.333 participants in each handedness category. A statistic used to determine whether the proportions differ from what we would expect were the null true is called the *chi-squared* (χ^2) statistic. This simply measures the discrepancy between observed frequencies in our sample and those expected if the null were true. The larger the difference the larger the chi-squared statistic becomes. Here is the formula:

Equation 8.1   General formula for the chi-squared test statistic.

$$\chi^2 = \sum_{i=1}^{k} \frac{(\text{observed}_i - \text{expected}_i)^2}{\text{expected}_i} \tag{8.1}$$

In words, this means that for each category i we calculate the squared discrepancies and divide each by their expected frequency. We should have learned by now that the reason for squaring the differences is that otherwise the sum would be zero. So for the categorical variable of type of handedness, we have counts in three categories, 348 for right, 42 for mixed and 28 for left-handed. All these terms are then added together.

With reference to our data above this gives us:

$$\chi^2 = \sum \frac{(\text{observed} - 139.333)^2}{139.333}$$

$$\chi^2 = \frac{(348 - 139.333)^2}{139.333} + \frac{(42 - 139.333)^2}{139.333} + \frac{(28 - 139.333)^2}{139.333}$$

$$= 312.501 + 67.994 + 88.960$$

$$= 469.455$$

Table 8.1 Statistical table for the χ^2 distribution for 1–12 degrees of freedom

df	.1	.05	.025	.02	.01	.005	.002	.001
1	2.71	3.84	5.02	5.41	6.63	7.88	9.55	10.83
2	4.61	5.99	7.38	7.82	9.21	10.60	12.43	13.82
3	6.25	7.81	9.35	9.84	11.34	12.84	14.80	16.27
4	7.78	9.49	11.14	11.67	13.28	14.86	16.92	18.47
5	9.24	11.07	12.83	13.39	15.09	16.75	18.91	20.52
6	10.64	12.59	14.45	15.03	16.81	18.55	20.79	22.46
7	12.02	14.07	16.01	16.62	18.48	20.28	22.60	24.32
8	13.36	15.51	17.53	18.17	20.09	21.95	24.35	26.12
9	14.68	16.92	19.02	19.68	21.67	23.59	26.06	27.88
10	15.99	18.31	20.48	21.16	23.21	25.19	27.72	29.59
11	17.28	19.68	21.92	22.62	24.72	26.76	29.35	31.26
12	18.55	21.03	23.34	24.05	26.22	28.30	30.96	32.91

The column for the .05 significance level has been shaded grey for convenience. More extensive tables can be found online, for example, http://www.medcalc.org/manual/chi-square-table.php.

The number of degrees of freedom associated with this statistic is one fewer than the number of categories, as we have two categories, $3 - 1 = 2$. The critical χ^2 with 1 degree of freedom for 5% significance is 5.99, which is the threshold that must be exceeded for statistical significance (Table 8.1). You can obtain critical χ^2 values in R by typing: *qchisq(0.950, 2)* which here will give the critical value for 5% significance and 2 degrees of freedom. (For 1 degree of freedom the critical value is 3.84, which is actually the same as the squared critical z value, 1.96^2.) Our calculated value of 469.5 is way higher than 5.99 so we can declare our finding as statistically significant. In other words, for our sample of 418 participants, it is highly unlikely that the apparent difference in proportions could have arisen by chance if the null hypothesis were true. The data in this analysis must be categorised counts (of people, choices, observations, etc.), and cannot be percentages, although it is acceptable to give the summary of your results in terms of percentages.

When this analysis is done in a statistical package we can obtain an exact probability. The R code and output for this analysis is given in Box 8.1. The p value given in the output is written as 2.2e-16, which is 2.2×10^{-16}. This is a very, very small number. It is a decimal point followed by 15 zeros before the first 2 (the probability is more than a billion times less likely than winning the lottery). So this is highly statistically significant. It is highly unlikely these proportions could have arisen if the 1:1:1 null hypothesis were true.

Box 8.1   R commands used to enter handedness data, do one-way chi-squared analysis and to obtain critical 5% significance value for 2 degrees of freedom. R code is proceeded by >, comments are given after #. Output from the commands appears below the code

```
> freq <- c(348,42,28)                          #enter data
> preference <- c("Right","Mixed","Left")
> handed <- data.frame(freq,preference)
> chisq.test(handed$freq)                        #the chi-squared test


     Chi-squared test for given probabilities


data:  handed$freq
X-squared = 469.4545, df = 2, p-value < 2.2e-16


> qchisq(0.950, 2)                               #critical value for 2 df
[1] 5.991465
```

One-way χ^2 results summary: In a sample of 418 patients with schizophrenia, there was very strong evidence for lateralised handedness, $\chi^2(2) = 469.45$, $p < .001$. Similar to those unaffected by schizophrenia, there was a large preponderance of right handers (83%), many fewer mixed handers (10%) and least of all, left handers (7%).

The bracketed 2 after the χ^2 signifies that we have 2 degrees of freedom.

Our example involved three participant categories. The number of categories can be any greater than 2 (with one we have nothing to compare!). The same analysis is done though the expected frequencies are calculated by dividing by the number of different choice categories available. Note: for expected values use as many digits after the decimal point as possible to increase the accuracy of the calculations. We can also use the test if we had specific a priori expectations about what the proportions would be according to some theory or previously collected data. For example, we may have learned that larger surveys using the same methodology in those unaffected by schizophrenia produced the following proportions of 85.2% for right, 8.5% for mixed and 6.3% for left-handed. We could then test to see whether our data differ from those, where we substitute the number corresponding to the proportions given as our expected values. Our expected values would then be

$$.852 * 418 = 356.136$$

$$.085 * 418 = 35.530$$

$$.063 * 418 = 26.334$$

$$\chi^2 = \frac{(348 - 356.136)^2}{356.136} + \frac{(42 - 35.53)^2}{35.530} + \frac{(28 - 26.334)^2}{26.334}$$

$$= 0.18586859 + 1.178184633 + 0.105398192$$

$$= 1.469451415$$

Again, we compare our obtained χ^2 value of 1.47 with the critical value for 2 degrees of freedom of 5.99. Our obtained value fails to reach (not even close) to the critical value, and we can therefore conclude that these proportions do not differ from those unaffected by schizophrenia, that is, patients with schizophrenia appear to have the same profile of handedness as those unaffected.

The same can be done to test the proportion of offspring phenotypes in breeding experiments where different proportions would be expected according to genetic theory, for example, Mendelian inheritance frequencies. For this reason the one-way chi-squared test is often referred to as the *goodness of fit* test since we are seeing how well our observed data fit with what the data would be according to specific criteria (the null hypothesis, a specific theory or previous data).

8.3 Two-way chi-squared

What we have done above is called a one-way chi-squared test because participants are categorised according to a single variable (handedness category). We can extend the test to include another variable and perform a two-way test. For example, the gender of the patients was also recorded. Table 8.2 shows the cross-tabulation, also known as a contingency table, of gender and handedness for our 418 patients with schizophrenia.

Here, we are interested to know whether men and women patients differ in their hand preferences. We need to calculate the expected frequencies in each of the cells. Let's think about the top right cell, which is for

Table 8.2 Contingency table for categorical variables handedness and gender in patients with schizophrenia

		Handedness			
		Right	Mixed	Left	Row total
Gender	Men	221 (228.95)	31 (27.63)	23 (18.42)	275
	Women	127 (119.05)	11 (14.37)	5 (9.58)	143
	Column total	348	42	28	418

Expected frequencies are shown in brackets. These values are given to two decimal places.

left-handed men. There are more men than women, in fact the probability of a randomly selected person being a man in this study is the number of men divided by the total number of patients, therefore: 275 / 418 = .657894737. The fewest number of patients are left-handed, and the corresponding probability of a randomly selected patient from this study being left-handed is the number of left handers divided by the total number of patients, therefore: 28 / 418 = .066985646. Now the probability of a patient being **both** a man and left-handed is these two probabilities multiplied together .657894737 ∗ .066985646 = .044069504. Therefore, out of 418 patients in the study we would expect .044069504 ∗ 418 = 18.42105263 men to be left-handed if they were as likely as the women to be left-handed. This compares with the 23 that actually were so in this study, so there is a small discrepancy.

A quicker way of doing the very same calculation is to use the *marginal* totals. The marginal totals are those in the margins (row and column totals) of the contingency table (Table 8.2). We just multiply the marginal totals that refer to each cell in the table and divide by the grand total (418). For example, the expected number of left-handed men would be (row total ∗ column total)/grand total, or (275 ∗ 28) / 418 = 18.42105263, giving us the same answer we calculated above. The expected frequency for right-handed men would be (275 ∗ 348) / 418 = 228.9473684. The expected frequencies for the remaining cells are similarly calculated using a calculator or Excel. The chi-squared statistic is now calculated from the general formula given in Equation 8.1 and using the values in Table 8.2:

$$\chi^2 = \sum \frac{(\text{observed} - \text{expected})^2}{\text{expected}}$$

$$= \frac{(221 - 228.9473684)^2}{228.9473684} + \frac{(31 - 27.63157895)^2}{27.63157895} + \frac{(23 - 18.42105263)^2}{18.42105263}$$

$$+ \frac{(127 - 119.0526316)^2}{119.0526316} + \frac{(11 - 14.36842105)^2}{14.36842105} + \frac{(5 - 9.578947368)^2}{9.578947368}$$

$$= 0.275874168 + 0.410626566 + 1.138195489 + 0.530527246 + 0.789666474$$

$$+ 2.188837478$$

$$= 5.333727422$$

The degrees of freedom for a 2 × 3 contingency table, such as in this example, are worked out by multiplying the number of rows − 1 by the number of columns − 1: (2 − 1) ∗ (3 − 1) = 2. Again the critical value to compare our calculated χ^2 value is 5.99. Our value of 5.33 is a bit less than the critical value, which indicates a marginally non-significant difference from the expected frequencies assuming our null hypothesis had been true. The R code and outputs are given in Box 8.2. Since 5.33 approaches the critical value of 5.99, the p value is also close to statistical significance at .069. The research paper DeLisi *et al.* (2002) included those unaffected by schizophrenia (non-patients). Table 8.3 gives the cross-tabulation.

Box 8.2   R commands used to do two-way chi-squared analyses

```
> M <- as.table(rbind(c(221, 31, 23), c(127, 11, 5)))   # patients schizophrenia data
> dimnames(M) <- list(Gender=c("Men","Women"), preference=c("Right","Mixed","Left"))
> M                                              # Prints cross-tabulation
      preference
Gender  Right Mixed Left
  Men     221    31   23
  Women   127    11    5
> (Xsq <- chisq.test(M,correct=FALSE))           # Prints test summary

        Pearson's Chi-squared test

data:  M
X-squared = 5.3337, df = 2, p-value = 0.06947

> Xsq$expected                                   # expected counts under the null
      preference
Gender     Right    Mixed       Left
  Men    228.9474 27.63158 18.421053
  Women  119.0526 14.36842  9.578947
>
> M <- as.table(rbind(c(109, 13, 13), c(136, 12, 5)))   #non-patients data
> dimnames(M) <- list(Gender=c("Men","Women"), preference=c("Right","Mixed","Left"))
> M                                              # Prints cross-tabulation
      preference
Gender  Right Mixed Left
  Men     109    13   13
  Women   136    12    5
> (Xsq <- chisq.test(M,correct=FALSE))           # Prints test summary


        Pearson's Chi-squared test

data:  M
X-squared = 5.4674, df = 2, p-value = 0.06498


>
> M <- as.table(rbind(c(348, 42, 28), c(245, 25, 18)))  #patients & non-patients data
> dimnames(M) <- list(Patients=c("Patient","Non-patient"),
preference=c("Right","Mixed","Left"))
> M                                              # Prints cross-tabulation
            preference
Patients      Right Mixed Left
  Patient       348    42   28
  Non-patient   245    25   18
> (Xsq <- chisq.test(M,correct=FALSE))           # Prints test summary


        Pearson's Chi-squared test

data:  M
X-squared = 0.4555, df = 2, p-value = 0.7963
```

Table 8.3 Contingency table for categorical variables handedness and gender in non-patients

		Handedness			
		Right	Mixed	Left	Row total
Gender	Men	109	13	13	135
	Women	136	12	5	153
	Column total	245	25	18	288

When a two-way chi-squared test is carried out on these data in the same way as we did for the patient-group we get a very similar result. The R code and outputs are also in Box 8.2, the second analysis shown there. Again, a marginally non-significant result with $p = .065$, and the proportions very similar, notably proportionally fewer women left handers. Perhaps we could combine the data? First we should check whether the overall proportions of handedness differ between patients with non-patients, ignoring gender. This is shown in the third analysis given in Box 8.2. With $p = .796$ there is no evidence of a difference here between patients and non-patients, although we need to be slightly cautious because the proportions of men and women in the patient and non-patient groups are different (275:143 and 135:153, respectively). So combining the data across the patient and non-patient groups we get Table 8.4.

Analysis of the combined patient and non-patient data to see whether there is an association between gender and handedness we obtain a $p = .0047$. The code and output is given in Box 8.3. This is now highly statistically significant. Looking at the cross-tabulation and clustered bar chart in Figure 8.1, we can see that there are proportionately fewer women (light bars) who are left-handed compared with men (dark bars). One way of determining which cells contribute most to the inflated χ^2 is to examine the standardised residuals (also known as Pearson residuals). These are calculated using the formula given in Equation 8.2.

Table 8.4 Contingency table for categorical variables handedness and gender, in patients and non-patients combined

		Handedness			
		Right	Mixed	Left	Row total
Gender	Men	330	44	36	410
	Women	263	23	10	296
	Column total	593	67	46	706

Box 8.3   R commands and outputs for the two-way chi-squared analyses involving combined data for patients and non-patients. This is the analysis performed for the final summary given in the text

```
> M <- as.table(rbind(c(330, 44, 36), c(263, 23, 10)))      #combined data
> dimnames(M) <- list(Gender=c("Male","Female"), preference=c("Right","Mixed","Left"))
> M                                              # Prints cross-tabulation
      preference
Gender   Right Mixed Left
  Male     330    44   36
  Female   263    23   10
> (Xsq <- chisq.test(M,correct=FALSE))           # Prints test summary

      Pearson's Chi-squared test

data:  M
X-squared = 10.7193, df = 2, p-value = 0.004703

> barplot(M,beside=TRUE,xlab="Hand Preference",ylab="Frequency") # Prints barplot
> Xsq$residuals                                  # Pearson residuals
      preference
Gender       Right       Mixed        Left
  Male   -0.7747194   0.8161057   1.7966594
  Female  0.9117809  -0.9604892  -2.1145201
```

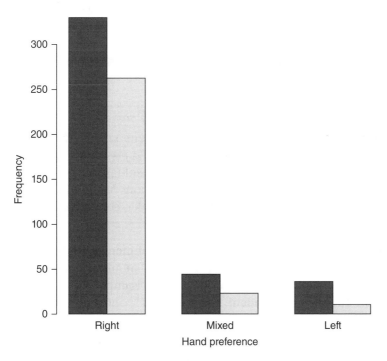

Figure 8.1 Clustered bar chart showing hand preferences according to gender. The dark bars represent men, and the light bars women. Proportionally fewer women are left-handed. The data show patient and non-patient groups combined together.

Equation 8.2 Formula for the calculation of standarised residuals in chi-squared.

$$\text{standardised residual} = \frac{(\text{observed} - \text{expected})}{\sqrt{\text{expected}}} \tag{8.2}$$

The residuals calculated indicate which cells are most different from what we would expect if the null hypothesis were true. They are used for easy detection of cells in the cross-tabulation rather than for formal statistical significance testing. The residuals can be used for significance testing, you can compare them with 2 (rough rule-of-thumb based on $z = 1.96$). However, the overall table will often be statistically significant while individual cells will not reach the critical value of 2. This is because the overall significance takes into account the whole pattern of discrepancies from the null hypothesis. Here are the residuals for our example:

	Standardised residuals		
	Right	Mixed	Left
Men	−0.77472	0.816106	1.796659
Women	0.911781	−0.96049	−2.11452

The largest absolute (ignoring negative signs) values were for left handedness of both men and women (highlighted: 1.80 and 2.11, respectively), confirming what we saw in Figure 8.1.

> **Two-way χ^2 results summary:** In a sample of 706 participants (410 men and 296 women, 418 patients and 288 non-patients) men and women differ significantly in hand preferences, $\chi^2(2) = 10.72, p < .005$. A clustered bar chart is given in Figure 8.1. A standardised residual analysis showed that men and women differed most in their proportions of left handedness. Proportionately fewer women (3%) were left-handed compared to men (9%).

In this example there are unequal numbers of men and women. It is not a necessary assumption that there should be equal numbers of men and women as the test is assessing the **relative proportions**. It is important that participants make their choices **independent of each other**[1], and that each participant

---

[1] It seems that there are few research reports without statistical flaws or which supply all the data necessary to replicate their analyses. The example used here for illustrative purposes was from DeLisi *et al.* (2002). It turns out that they recruited 418 patients from 259 families. Since some of the patients were from the same families this violates the independence assumption and somewhat compromises their conclusions.

Table 8.5   In a fictional study of colour preference in 15 boys and girls. Two of the cells have **expected** frequencies less than 5

		Preferred colour		
		Blue	Pink	
Gender	Boys	7	0	7
	Girls	4	4	8
		11	4	15

Table 8.6   The cell frequencies in Table 8.5 are replaced by letters, used in the formula for Fisher's Exact test

	Preferred colour		
	Blue	Pink	
Male	a	b	$a + b$
Female	c	d	$c + d$
	$a + c$	$b + d$	$a + b + c + d$

contributes only once (so total number across the different categories must add to the total number of participants in our sample). The numbers of counts **cannot be percentages**, they must be the categorised counts of participants (or observation counts). For the chi-squared analysis to work accurately it is necessary that **no more than 20% of the cells in a table have expected frequencies less than 5**. For our 2×2 example we could tolerate no more than 1 cell having an expected frequency less than 5. If there are too many cells with frequencies less than 5, then some levels of one of the variables could be combined together in some meaningful way to satisfy this requirement.

If we had a 2×2 cross-tabulation with inadequate expected frequencies, then a valid analysis can be done using Fisher's Exact test. This test is independent of the chi-squared distribution. It will only work on a 2×2 design and not on any larger cross-tabulation. Consider the data shown in Table 8.5.

Table 8.5 can be represented by Table 8.6 where numbers are replaced by letters.

The following (rather surprising!) formula can then be used to calculate an exact p value (the exclamation mark in the formula means *factorial*, and is explained in the Basic Maths for Stats Revision section):

$$p = \frac{(a+b)!*(c+d)!*(a+c)!*(b+d)!}{(a+b+c+d)!*a!*b!*c!*d!}$$

$$= \frac{(7+0)!*(4+4)!*(7+4)!*(0+4)!}{(7+0+4+4)!*7!*0!*4!*4!}$$

$$= \frac{7!*8!*11!*4!}{15!*7!*0!*4!*4!}$$

$= .051$ (this is the one-tailed test p value)

Fortunately, it is available in most statistical packages. There are two issues that arise using this test. One is that, strictly speaking, it only applies to a situation where we have fixed marginal totals. In our example, we would have to suppose that in a replication of our study the same numbers of boys and girls were recruited (7 and 8, respectively) and also that the same totals for blue and pink colour preference (11 and 4, respectively). This would still allow the four cells within the table to vary. Whether this restriction makes a huge difference is debatable, and most people are happy to use it on data where one or both marginal totals are random (free to vary). The other issue is whether a one- or two-tailed p value should be used. This should be decided in advance of collecting the data (if you can remember to do so!). Generally though, as suggested elsewhere, it is better to use two-tailed testing. This is a more conservative approach (less powerful), but at least it does not leave you open to accusations of using a one-tailed test simply to achieve statistical significance.

This chi-squared test is known as a test for *association* (e.g. 'Is the pattern of hand preferences associated with gender'? as alternative hypothesis) or as a test for *independence* (e.g. 'Is the pattern of hand preference independent of gender'? as null hypothesis). Both names and phrasings are equally valid. The cross classification consists of two variables in our example, one with two levels and the other with three levels (2×3). We could have any number of levels for each of the two variables. The chi-squared test cannot be used to analyse more than two variables/dimensions. For example, in the 2×3 design described above, we could not add another variable such as age: Young/Old as this would make it a 3-dimensional $2 \times 3 \times 2$ design. Such a design can be visualised by two 2×3 contingency tables, one for hand preference and gender in young people, and a second for hand preference and gender in old people. Analysis of such categorical data consisting of more than two dimensions requires the use of an advanced statistical technique called *log-linear analysis*.

8.4   The odds ratio

In many medical contexts we have patients who are either given a treatment or a placebo, and we record those who lived/died (or improved/remained ill, etc.) where *risk* is relevant. Here, we have a 2×2 cross-tabulation. The *odds ratio* is a statistic which allows us to determine whether an intervention either improves or worsens a particular condition. We first calculate the odds for people receiving the intervention: affected/unaffected by the condition. We do the same for those not receiving the intervention. We then calculate the ratio of those odds. If the intervention improves the condition then we would expect its odds to be lower than the odds for those not receiving the intervention. Hence we would expect the OR, which is the ratio of the intervention odds to the non-intervention odds, to be small and less than 1. Conversely, if the intervention exacerbates/worsens the condition, then the odds ratio will be greater than 1. If there is no effect of the intervention, then the odds ratio will be close to 1 – this is the null hypothesis. As an example, we will use data from a well-known study that examined the effect of folic acid supplements on neural tube defects. Neural tube defects can be a serious and debilitating defect in the development of the spinal cord and brain (and include spina bifida). Neural tube defects occur in about 1 in 1000 live births. The study was carried out with women who had been affected by neural tube defects in a previous pregnancy. Because the association looked so strong early on in the trial, it had to be halted, and all participants given folic acid supplements. Participating women were randomly assigned to the different treatment groups. The actual study was slightly more complicated with some groups receiving other vitamins, or combinations of vitamins and folic acid. However, the relevant data are shown cross-tabulated in Table 8.7.

Across both treatment groups, 2.3% (27 / 1195) had neural tube defects. In the no-treatment group, 3.5% (21 / 602) of babies had neural tube defects. There are almost equal numbers in the two treatment groups, but note that 3.5 (21 / 6) times the number of babies born to women not receiving treatment had defects compared with those born to women receiving folic acid.

The compact and simple odds ratio expresses this finding. First, we need to calculate the odds of having defects for each treatment, and then calculate the ratio of those odds; hence, 6 / 587 = 0.010221465 and 21 / 581 = 0.036144578. The ratio of these odds 0.010221465 / 0.036144578 gives us 0.282793867. Rounding to 3 decimal places we have 0.283. This means that the babies of women treated with folic acid had a probability of less than one third of the no-treatment group of having neural tube defects. Folic acid supplements appear to be protective in babies of these women. The inverse of this odds

Table 8.7 Contingency table for categorical variables: type of intervention (4 mg/day folic acid supplements versus no supplement) and whether or not the baby had neural tube defects

		Babies with neural tube defects		
		Yes	No	
Treatment	Folic acid	6	587	593
	None	21	581	602
		27	1168	1195

ratio is $1 / 0.282793867 = 3.536144578$, and can also be useful in thinking about the effect. Here, we can say that babies in those mothers receiving no treatment were about 3.5 times more likely of developing neural tube defects than the folic acid group.

It is important to remember that one is actually only talking about a relatively small percentage (2%) of babies that develop neural tube defects, although this is a disabling and life-threatening condition for those babies affected. This is a relative statistic for those mothers whose babies developed the defects. Most babies were unaffected. In absolute risk terms, we can think of folic acid reducing the incidence of neural tube defects from 0.034883721 ($21 / 602$) to 0.010118044 ($6 / 593$). The absolute risk reduction is therefore the difference between these values, 0.024765677 (about 2.5%), and used below to calculate the *number needed to treat* statistic. In other words, the folic acid resulted in nearly 3 fewer babies per 100 babies having neural tube defects than we would expect without folic acid. Is this statistically significant?

Like chi-squared the statistical significance of the odds ratio is sensitive to the total number of participants in the study. We can calculate the statistical significance as we would for a 2×2 contingency table. The R code for chi-squared and 95% confidence interval, and respective outputs are given in Box 8.4. This shows that folic acid significantly reduced neural tube defects in babies, $\chi^2(1) = 8.30, p = .004$.

Box 8.4 *R commands and outputs for the chi-squared test, odds ratio analysis and odds ratio 95% confidence interval. As before code is given after the prompt >, except for the function which occurs between {} and code is indented*

```
> M <- as.table(rbind(c(6, 587), c(21,581)))        #4 mg/d folic acid, Lancet study 1991
> dimnames(M) <- list(Treatment=c("Folic acid","None"),Outcome=c("Neural Tube
Deficit","Healthy"))
> M                                                 # Prints cross-tabulation
          Outcome
Treatment    Neural Tube Deficit Healthy
Folic acid                     6     587
None                          21     581
> (Xsq <- chisq.test(M,correct=FALSE))              # Prints test summary


      Pearson's Chi-squared test

data:  M
X-squared = 8.2968, df = 1, p-value = 0.003971

> barplot(M,beside=TRUE,xlab="Outcome",ylab="Frequency")    # Prints barplot, useful?


> # Odds ratio with confidence interval
> oddsCI <- function(table,level)                   # level is %, e.g. 95
{
    a <- (((100-level)/2)+level)/100
    z <- qnorm(a)
```

```
    SEL <- sqrt(1/table[1,1] + 1/table[1,2] + 1/table[2,1] + 1/table[2,2])
    limit <- z*SEL
    rat1 <- table[1,1]/table[1,2]
    rat2 <- table[2,1]/table[2,2]
    OR <- rat1/rat2
    LOR <- log(OR)
    Lupper <- LOR + limit
    Llower <- LOR - limit
    upper <- exp(Lupper)
    lower <- exp(Llower)
    cat("Odds ratio = ",OR,"\n",level,"% limits:",lower," ",upper,"\n")
    }
> oddsCI(M,95)                                # for 95% CI
Odds ratio =  0.2827939
95 % limits: 0.1133193    0.7057258
```

We can work out the 95% confidence interval for the odds ratio by first calculating it for the natural log of the odds ratio, abbreviated to ln(odds ratio). To obtain the natural log in Excel use the command =*LN()* with the number to be logged within the brackets. (Taking the antilog will return the odds ratio.) The standard error (SE) of ln(odds ratio) is simply calculated by taking the square root of the sum of the inverse frequencies.

Step 1. Calculate the square root of the sum of the inverse frequencies:

$$SE(\ln(\text{odds ratio}))$$

$$= \sqrt{\frac{1}{6} + \frac{1}{587} + \frac{1}{21} + \frac{1}{581}}$$

$$= \sqrt{0.166666667 + 0.001703578 + 0.047619048 + 0.00172117}$$

$$= \sqrt{0.217710462}$$

$$= 0.466594537$$

Step 2. The odds ratio for Table 8.7, calculated a few paragraphs above, was 0.282793867. The ln(odds ratio) is $\ln(0.282793867) = -1.263037031$.

In Excel use =*LN(0.282793867)* or on a calculator use the *ln* button.

Step 3. We need to multiply the SE obtained in Step 1 by the critical z value for 5% significance, then add and subtract this to the ln(odds ratio).

$$-1.263037031 \pm (1.96 * 0.466594537) \text{ gives:}$$

$$-2.17754552 \text{ and } -0.348528543$$

Step 4. Antilogging these values $(=EXP(-2.17754552)$ and $=EXP(-0.348528543))$ we get 0.11331933 and 0.705725771. These limits do not include the null hypothesis value of 1, hence we can conclude that our obtained value of 0.282793867 is statistically significant at the 5% level (which we already knew from the chi-squared test).

Limits calculated for the reciprocal odds ratio of 3.536144578, calculated by reciprocals of 0.11331933 and 0.705725771 (1 / 0.11331933 and 1 / 0.705725771), to give us the corresponding 95% limits above 1 (8.824619801 and 1.416980987, respectively). Note that the neither of the odds ratios are in the centre of their respective 95% confidence intervals. This is because they originated from a logarithmic scale (where they are symmetrical, but lose their symmetry when converted back to the odds ratio).

> **Odd ratio results summary:** In a study of 1195 women an odds ratio of 0.28 (95% confidence interval of 0.113 to 0.706) was found in a study on the administration of folic acid and the occurrence of neural tube defects. Folic acid supplements given to mothers significantly reduced neural tube defects in their babies, $\chi^2(1) = 8.30, p = .004$.

It is worth considering this study a bit further. The study was carried out in women whose previous pregnancy produced an infant or foetus affected by neural tube defects. So these were not randomly selected pregnant women, but were an unusual subgroup who had previous medical problems. The study was important in driving health policy for prophylactic folic acid by all women – yet most of these women would not be vulnerable to their babies developing the condition. There is evidence that excess folic acid is associated with certain health risks suggesting that there should be a careful risk–benefit analysis for giving supplements and fortified food (in the United States, e.g. many grain products are fortified).

The odds ratio is often used (some would say overused) in epidemiology, particularly in cohort and case-control studies. A value less than 1 means a reduction in mortality or morbidity, while a value greater than 1 means an adverse effect. In the form of ln(odds ratio) it is also used in logistic regression where the outcome is binary such as live/die, healthy/ill, improve/not improve. If the risk is very small then the odds ratio approximates another useful medical statistic known as *relative risk*. In our example the relative risk is (6 / 593) / (21 / 602) = 0.29, which compares favourably with our odds ratio of 0.28. As an effect size, an odds ratio between 0.5 and 2.0 is not worth considering clinically important. Few if any epidemiological studies with such modest ratios have been replicated in randomised control trials (RCTs). So try your best to ignore those screaming media headlines saying things like 'Chocolate may protect the brain and heart' http://www.bbc.co.uk/news/health-14679497 or 'Red meat increases death, cancer and heart risk' http://www.bbc.co.uk/news/health-17345967 and their misleading reporting (see Penston, 2010).

We can also calculate the medically relevant *number needed to treat* (NNT) statistic, which is the reciprocal of the absolute risk reduction (we noted above it was about 2.5%). This tells us how many patients need to be treated to reduce mortality/morbidity by 1 person. In our example this works out to be 1 / 0.024765677 = 40.37846397. This means we need to treat (give supplements to) 40 at-risk pregnant women (who previously had medical problems) in order to prevent one of their babies developing neural tube defects. In other words, if treatment were **not** given, then one additional baby would develop neural defects in addition to the number expected for that population of women. In terms of the number of neural tube defective babies per 100, this would mean increasing from 1 baby (100 ∗ 6 / 593) to 3.5 (100 ∗ 21 / 602) babies if folic supplements were not used. You can see that the recommendation for the use of supplements for those women at risk is fairly compelling.

In therapeutic terms (if the intervention is beneficial), the ideal NNT is 1, and an NNT of 100 is poor. The particular NNT value obtained will depend on the context (e.g. seriousness of the disease, and the cost and possible adverse side effects of treatment). A risk–benefit analysis should always be considered, particularly when the NNT is high. As noted above, the use of relative statistics, like the odds ratio and relative risk, can be misleading. A more transparent way of presenting the data would involve presenting the absolute risk reduction (e.g. the reduction from 3.3 babies to 1 baby for every 100 at-risk mothers administered folic acid supplements).

8.5 Summary

- Categorical data consist of data in the form of frequencies of observations in different categories (or classes).

- For simple comparison of frequencies across a set of categories, the one-way chi-squared test is used. Typically the null hypothesis is that we expect equal frequencies across the categories, although specific expected frequencies can be defined according to theory.

- For cross-tabulation of two variables, the two-way chi-square test is used. The null hypothesis for these data is that the two variables are independent of each other. It is also known as a test of association.

- The calculated chi-squared value χ^2 can be compared against a tabulated critical value according to the degrees of freedom of the analysis. Alternatively, Excel or R can be used to either obtain the critical value or to obtain the exact *p* value. For a two-way analysis, examination of the standardised

(Pearson) residuals allows one to determine where in the cross-tabulation the largest effects are.

- For the analysis to work accurately, no more than 20% of the cells in a table should have expected frequencies less than 5.

- A 2 × 2 cross-tabulation with insufficient expected frequencies can be analysed using Fisher's Exact test. This test is available in most statistical packages.

- The odds ratio is widely used in a 2 × 2 cross-tabulation, typically where one variable consists of whether or not treatment was given, and the other variable consists of whether or not a change in outcome occurred. This is a common format in medical epidemiological studies, treatment (Yes/No) and outcome (Better/Not Better). The odds ratio expresses in a single number the relationship between treatment and outcome. The null hypothesis is that there is no relationship between these variables, and therefore equal to 1. Care should be taken in its use since it is a relative, rather than absolute, statistic. 95% confidence intervals can be calculated on the odds ratio.

References

DeLisi, L.E., Svetina, C., Razi, K., Shields, G., Wellman, N. and Crow, T.J. (2002) Hand preference and hand skill in families with schizophrenia. *Laterality*, 7(4):321–332.

MRC Vitamin Study Research Group. (1991) Prevention of neural tube defects: results of the Medical Research Council Vitamin Study. *The Lancet*, 338:131–137.

Penston, J. (2010) *Stats.con: How We've Been Fooled by Statistics-Based Research in Medicine.* The London Press.

9

Non-Parametric Tests – 'An Alternative to other Alternatives'

9.1   Aims

This chapter focuses on alternative statistical procedures which have been popular because no, or few, assumptions need to be made about the data. We will look in turn at non-parametric equivalents to the different types of parametric tests already covered in Chapter 6. Non-parametric tests are often referred to as *robust* procedures as they tend to be unaffected by extreme outlier data points, but they are generally not as powerful as parametric tests on normally distributed data. As we will see at the end of the chapter, these procedures are very good in their limited way, but lack the sophistication and flexibility of their parametric counterparts.

9.2   Introduction

In many of the analyses referred to in earlier chapters we were interested in estimating one or more parameters from population distributions. We also had to make certain assumptions about the populations from which the data came. For example, in Chapter 6, we had to assume that our collected data were normally distributed and that the variances in the different groups were approximately equal. We had to estimate both the population mean(s) and variance(s). Analyses that involve making specific assumptions and estimating parameters are known as *parametric* tests. If we are unable to make these

Starting Out in Statistics: An Introduction for Students of Human Health, Disease, and Psychology
First Edition. Patricia de Winter and Peter M. B. Cahusac.
© 2014 John Wiley & Sons, Ltd. Published 2014 by John Wiley & Sons, Ltd.
Companion Website: www.wiley.com/go/deWinter/startingstatistics

assumptions, or analyses do not depend on estimating specific parameters, then there are alternative statistical analyses known as *non-parametric* tests. They are also referred to as *distribution-free* tests because no assumptions need to be made about the distribution(s) of the population(s) from which the data were obtained. The analyses of categorical (nominal) data using chi-squared tests (Chapter 8) are also non-parametric. It is worth remembering that non-parametric tests are not completely assumption free. We still need to assume that samples are selected at random and, if comparisons are made between groups, that they are randomly assigned and independent of each other.

In general, parametric tests are quite robust to violations of assumptions such as normality and heterogeneity of variances. Usually we can't know whether the population(s) from which we sample is/are normally distributed because we only sample once. So we are forced to examine our particular sample(s) for normality. If there are outliers in two or more samples, then that would affect both the normality and homogeneity of variance assumptions. If the data are positively skewed, it may be possible to normalise the data with a transform, such as a log transform (see Chapter 6). This often also cures the heterogeneity of variances problem (two for the price of one). You can then proceed with a parametric analysis remembering to transform back to the original scale of the measurements. In larger samples, say greater than 30 per group, the sampling distributions become normal and hence parametric tests are likely to be robust against assumption violations. However, if there are small samples, say less than 30 per group, and the data look non-normal or come from a population known to be non-normal (e.g. waiting times for buses), it would be advisable to use non-parametric statistical tests. Non-parametric tests are also useful when dealing with noisy data, and data that include errors or outliers which lack the necessary justification for removal. Looking across the groups, if the largest variance is more than four times the smallest variance, and there are large differences in sample sizes (say 2:1), then it's strongly advisable to use a non-parametric approach. Note that if there are large differences in variances, then that too is an interesting finding which indicates that an intervention or treatment affects the variability of the data, and should be reported along with the difference of location (mean or median). Finally, when ordinal (ranked order) data are collected it is often more appropriate to examine medians rather than means, and here non-parametric tests are often preferred. Medians, rather than means, should be reported (and plotted) for non-parametric analyses.

Apart from the chi-squared test and Spearman's rank correlation, we have already encountered a non-parametric test in Chapters 2 and 3, where we had 10 patients, half of whom received treatment and the other half a placebo. The probability calculation was based on how many ways there were of selecting the data to get the obtained result (1) and the total number of ways of selecting the data (252). This is known as a permutation or randomisation test and such tests are given greater attention in the next chapter.

A suitable reference giving more information on the tests described here is Sprent (1992).

9.3   One sample sign test

Some explorers have returned from the deep jungle to an outlying settlement. Curiously, but maybe not surprisingly, most of them complain of weakness and infections of one form or another. A quick blood sample is taken from each person and, among other measurements, a white cell count done. Unfortunately, the makeshift lab results have poor reliability, but they are known not to be consistently biased either low or high. Because the data are noisy with some outliers it would be sensible to use a non-parametric test. We want to compare their values with the white blood cell count in healthy individuals which is 7 $(\times 10^9/L)$[1]. Values for our explorers, indicating in the bottom row whether they are above (+) or below (−) the healthy value, are given in Table 9.1.

Table 9.1   Raw data from 10 explorers giving white blood cell (WBC) counts

Explorer	1	2	3	4	5	6	7	8	9	10
WBC $\times 10^9/L$	2	11	3	6	6	1	1	5	4	5
Difference from healthy	−	+	−	−	−	−	−	−	−	−

The bottom row indicates whether each explorer's count was above (+) or below (−) the healthy value of $7 \times 10^9/L$.

Only one of these readings exceeds the healthy value, all the others are lower, so it looks like their white cell counts are on the low side. According to the null hypothesis of no difference from healthy we would expect equal numbers of values to be above as below the value of 7 $(\times 10^9/L)$. So the probability of a + or a − will be equal. We could denote each of the probabilities of + and − separately, as p and q, respectively. In our case our null hypothesis is that they are equal: $p = q = .5$. These probabilities must add up to 1. We need to determine the probability of obtaining just one reading above the healthy value (which is the same as calculating the probability of obtaining nine readings below healthy).

The results of the calculations, using Excel, are reported in Table 9.2. Most statistical packages will calculate a sign (or binomial) test if you enter the data in binary form, for example, 1 and 0 rather than the + and − signs we have

---

[1] Actually the normal acceptable range is 4–11 $\times 10^9/L$, so choosing the specific value of $7 \times 10^9/L$ is convenient here for illustrative purposes only.

Table 9.2 Calculations for the sign test using the binomial distribution (see text)

Number of + signs	0	1	2	3	4	5	6	7	8	9	10
Combinations	1	10	45	120	210	252	210	120	45	10	1
Probability	.0010	.0098	.0439	.1172	.2051	.2461	.2051	.1172	.0439	.0098	.0010
Cumulative Prob.	.0010	.0107	.0547	.1719	.3770	.6230	.8281	.9453	.9893	.9990	1.0000

The cumulative probabilities may appear not to add due to rounding to only four decimal places in the probability figures shown.

used here, depending on whether or not the reading exceeds a specified (for our example, the healthy) value. We calculate the number of combinations (second row) for each possible outcome from 0 to 10 plus signs. Here $n = 10$ and each number of plus signs is referred to as r, and we use the binomial coefficient ^nC_r (which we already encountered in Chapter 3).

$$^nC_r = \frac{n!}{r! * (n - r)!}$$

So to work out the number of combinations of just no + sign would be

$$^{10}C_0 = \frac{10!}{0! * (10 - 0)!}$$

Where by definition 0! is 1 gives us:

$$= \frac{10!}{10!} = 1$$

Doing the same for one + sign would be

$$^{10}C_1 = \frac{10!}{1! * (10 - 1)!}$$

$$= \frac{10!}{9!} = 10$$

And so on up to 10 + signs. The probability (third row) of each of these outcomes is then obtained by multiplying the combinations by $p^r * q^{n-r}$. Hence, the probability of getting zero + signs:

$$p(0) = 1 * p^0 * q^{10}$$

$$= 1 * 0.5^0 * 0.5^{10} = 1 * 1 * 0.0009765625 = .0009765625$$

And for one + sign (where there were 10 combinations from our binomial coefficient):

$$p(1) = 1 * p^1 * q^9$$
$$= 10 * 0.5^1 * 0.5^9 = .009765625$$

The bottom row is the cumulative probability, adding each value from the row above to the previous, as we go from 0 to 10. The final value is 1, accounting for all possible outcomes. The shaded column shows the relevant calculations for our study where we had just one + sign.

The cumulative probability is .0107. This value is the sum of the tail .0010 + .0098. Since we should allow for the possibility that our explorers had higher white blood cell counts than healthy individuals, we need to do a two-tailed test which requires us to multiply our cumulative probability value by 2: .0107 * 2 = .021. Since this value is less than .05 we can say that the low white blood cell count in these explorers ($n = 10$) was statistically significant. The median white blood cell count should be calculated and given in the summary. This and the binomial calculation in R are given in Box 9.1.

Box 9.1   *R code for performing the binomial test and obtaining the median value from the data. Comments are in uppercase or after #. R code occurs after each prompt >*

```
X IS THE NUMBER OF + VALUES, N IS SAMPLE SIZE, P IS PROBABILITY OF EACH TYPE OF OUTCOME


> binom.test(x=1, n=10, p=.5)


Exact binomial test

data:  1 and 10
number of successes = 1, number of trials = 10, p-value = 0.02148
alternative hypothesis: true probability of success is not equal
to 0.5
95 percent confidence interval:
 0.002528579 0.445016117
sample estimates:
probability of success
                0.1


THE 95% CONFIDENCE INTERVAL IS FOR THE PROBABILITY OF A + VALUE, BEST ESTIMATE IS .1


> median(c(2,6,3,11,6,1,1,5,4,5))    #calculates the median
[1] 4.5
```

Results summary: We can conclude from a sign test that the white blood cell count (median $= 4.5 \times 10^9$/L) from 10 explorers was significantly lower than normal, $p = .021$. The data are shown in a boxplot with superimposed dotplot in Figure 9.1.

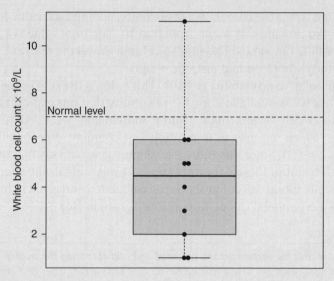

Figure 9.1 Boxplot with superimposed dotplot showing individual data points (filled circles). All but one of the 10 white blood cell counts fall below the normal level of 7×10^9/L (reference shown by dark grey dashed line). The median count is indicated by the thick line within the box.

It is worth considering the limitations of this study. What was our population? Presumably it would be all conceivable healthy explorers going from civilisation into this jungle area. However, this was not a random sample, nor are the individuals independent of each other (they are a group). Perhaps their white blood cell counts were all on the low side before they went into the jungle. Our conclusions about causality need to be qualified – in other words, it may actually have nothing to do with the jungle and everything to do with this particular group of explorers (more blood tests needed!).

The sign test is a useful 'quick' test but it should be kept in mind that because only binary data (yes/no, 1/0, +/−) are used, the magnitudes of individual data points are lost. In our example we cannot tell how much above or below the normal value each individual's white blood cell count is, only that it is high or low. In the tests below these magnitudes are taken into account, and use rankings of the magnitudes.

9.4 Non-parametric equivalents to parametric tests

Each of the tests described in the sections below are the non-parametric equivalents to the parametric tests we have already looked at in Chapter 6. Table 9.3 illustrates the correspondences between the tests. As mentioned in Chapter 2, there are two general designs. Between-participant designs use independent groups of observations, while within-participant designs use groups of observations that are related to each other (either because repeated measurements have been done in different treatment conditions or there are matched cases in the different treatment conditions).

Table 9.3 Designs are either between participants (independent samples) or within participants (related samples)

Design	Between participants		Within participants	
Number of treatments	Two treatments	More than three treatments	Two treatments	More than three treatments
Non-parametric test	Mann–Whitney test	Kruskal–Wallis one-way ANOVA	Wilcoxon signed-rank test	Friedman test
Parametric test	Independent samples t-test	One-way independent samples ANOVA	Related samples t-test	Repeated measures ANOVA

The number of treatments (groups, conditions or samples) is given in the second row. The name of the non-parametric test is given in the third row, and the corresponding parametric test in the bottom row.

9.5 Two independent samples

We have a new treatment for spots to test, crocus flowers. We have managed to recruit 19 spotty-faced people to test out the new therapy. Using random assignment we allocate equal numbers to each of the groups, except that having an odd number, one of the groups (at random) has an extra participant. After 2 weeks of treatment the number of spots on the participants' faces is counted. A double-blind procedure is used to minimise a placebo effect and bias. The data are given in Table 9.4.

Why should we use a non-parametric test here? The crocus group has one whopping outlier. This may have arisen for a number of reasons: this was a very spotty person even from the start, the person may have had an adverse response to the crocus treatment, it may be a recording error, etc. Without the necessary justification for removing the outlier, it is best to perform a *Mann–Whitney test* which compares two independent groups. The calculations for the test are very straightforward and can be done by hand. The spot counts

Table 9.4   Raw data for study of the effect of crocus flowers on facial spots

Crocus	Placebo
0	3
1	2
60	1
0	4
2	5
0	3
0	6
0	2
2	1
	4

The left column gives the treatment (crocus) and the right column gives placebo (liniment which was identical to crocus preparation but lacked crocus ingredient).

are pooled together and ranked irrespective of their group, starting with the lowest values (0) and working up to the highest value (60) which is assigned a rank of 19. Since there are five zeros, we need to assign the same average rank to these tied values. Adding the first five ranks together gives $1 + 2 + 3 + 4 + 5 = 15$. Since there are five of them, we need to divide this value by 5 to obtain the mean rank, which gives us 3, so the five values of 0 are each assigned this rank. The same principle is used throughout the ranking of the spot counts. Hence the next ranking is 7 for the three values of 1 (one in the crocus treatment and two in the placebo), since counting 3 ranks above 5 the next in the series are 6, 7, 8. The mean rank is in the middle, 7. There are four values of 2. The ranks for them are 9, 10, 11, 12. The mean rank of these is 10.5, and so on. This is done all the way up to the last rank which is 19, as seen in Table 9.5.

The total and mean ranks are located at the bottom of each **Rank** column, from which we can see that the lowest ranks tend to be in the crocus group (i.e. members of the crocus group tend to have the least number of spots). The Mann–Whitney test statistic, variably called U or W can be calculated from the smaller total, 62, using the following formula, where n_1 is the number of observations in the first group:

$$U_1 = \text{Total rank} - \left(\frac{1}{2} * n_1 * (n_1 + 1) \right)$$
$$= 62 - (.5 * 9 * 10)$$
$$= 17$$

Table 9.5 Ranks added to the data shown in Table 9.4, with each set of ranks to the right of the original data columns

Crocus	Rank	Placebo	Rank
0	3	3	13.5
1	7	2	10.5
60	19	1	7
0	3	4	15.5
2	10.5	5	17
0	3	3	13.5
0	3	6	18
0	3	2	10.5
2	10.5	1	7
		4	15.5
Total rank =	**62**	Total rank =	**128**
Mean rank =	**6.89**	Mean rank =	**12.8**

The grey highlighted ranks are those which are tied (see text).

U_1 can be used to calculate the complementary U value, called U_2, using this formula, where n_2 is the number of observations in the second group:

$$U_2 = (n_1 * n_2) - U_1$$
$$= (9 * 10) - 17$$
$$= 73$$

The smaller of the two U values is selected and reported for the analysis. In former times this value would be looked up by poring over extensive tables stretching over many pages. When the smaller (if there is a smaller) sample size is >20, then a z value can be calculated using the following formula using the smaller U:

$$z = \frac{\left(U + \frac{1}{2} - \left(\frac{1}{2} * n_1 * n_2\right)\right)}{\sqrt{[n_1 * n_2 * (n_1 + n_2 + 1) / 12]}}$$

If there are ties then the formula is slightly(!) different, numerator simpler but denominator not:

$$z = \frac{\left(U - \left(\frac{1}{2} * n_1 * n_2\right)\right)}{\sqrt{[(n_1 * n_2 * (n_1 + n_2 + 1) / 12) - (n_1 * n_2 * \sum (d_i^3 - d) / [12 * (n_1 + n_2) * (n_1 + n_2 - 1)])]}}$$

where d_i is the number of tied values in the ith group of tied scores. For example, the first tied group consists of five tied zeros, so $d_1{}^3 - d_1 = 5^3 - 5 = 125 - 5 = 120$. The second group consists of three tied ones, so $d_2{}^3 - d_2 = 3^3 - 3 = 27 - 3 = 24$, and so on. The sum total of these tied calculations would be 216. The correction made to the z value by taking ties into account is quite small and many people ignore using the correction.

The z value can then be compared with 1.96. If it exceeds 1.96, then the result would be reported as statistically significant at .05 level.

Fortunately, all statistical packages will run the Mann–Whitney test for independent samples, though it is given other names like Wilcoxon–Mann–Whitney test, Mann–Whitney U test, or Wilcoxon's rank sum test (which is confusingly similar to the name for the paired test below) for independent samples. The R code and output are given in Box 9.2. At the end of the output is a warning message saying that the exact p value cannot be calculated because of ties in the ranks, which can be ignored as it generally makes little difference.

Box 9.2   R code for entering data, obtaining medians and sample sizes, and running the Mann–Whitney test for two independent samples

```
> data   <- c(0,1,60,0,2,0,0,0,2,3,2,1,4,5,3,6,2,1,4)   #enter data
> treat <- c(rep("crocus",9),rep("placebo",10))
> crocus <- data.frame(data,treat)
> tapply(data,treat,median)                        #gives medians
 crocus placebo
      0       3
> tapply(data,treat,length)                    #gives sample sizes
 crocus placebo
      9      10
> wilcox.test(data~treat,paired=FALSE,conf.int=FALSE) #wilcoxon (aka Mann-Whitney)
       Wilcoxon rank sum test with continuity correction

data:  data by treat
W = 17, p-value = 0.02251
alternative hypothesis: true location shift is not equal to 0

Warning message:
In wilcox.test.default(x = c(0, 1, 60, 0, 2, 0, 0, 0, 2), y = c(3,  :
   cannot compute exact p-value with ties
```

Results summary: We can conclude our analysis by saying that a Mann–Whitney test showed that the crocus treatment resulted in statistically significantly fewer facial spots compared to placebo, crocus median $= 0, n = 9$, placebo median $= 3$, $n = 10$, $U = 17$, $p = .022$. The data are plotted in Figure 9.2.

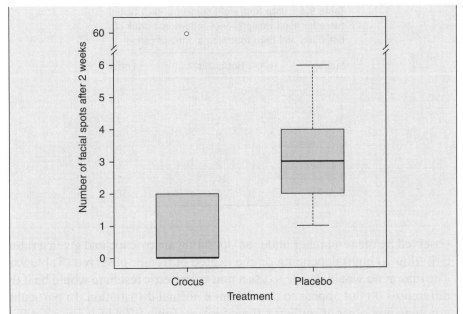

Figure 9.2 Boxplot of the effect of crocus flowers versus placebo on number of facial spots. The vertical axis has a break so that the extreme outlier at 60 crocus group is included – this allows easier comparison of the medians.

Had we used a t-test we would have obtained $t(8) = 0.622$, $p = .551$, not statistically significant. The analysis was affected by the very large outlier in the crocus treatment. It is notable that the order of the means is reversed from the medians (crocus mean = 7.22, placebo mean = 3.10), the crocus mean being strongly influenced by the outlier.

9.6 Paired samples

If we have a sample of participants we can test them before and after an intervention to see if the intervention has an effect. Using the same participants is helpful in that each participant acts as their own control. We plan a study to determine whether caffeine affects the degree of hand tremors in qualified surgeons. Obviously, shaky hands during surgery is something for us all to be concerned about. Eight surgeons were selected at random from medical staff from a number of hospitals in the region. On different days they were given either hot water or coffee to drink. The order of taking these different drinks was counterbalanced. After half an hour, their hands were

Table 9.6 Data from eight surgeons, each tested twice for hand tremors, once following a drink of hot water and once following a drink of coffee

Surgeon	Hot water	Coffee
1	2	8
2	3	4
3	1	7
4	0	9
5	4	10
6	2	6
7	10	1
8	1	10

observed by the researcher under an operating microscope and given a rating 0–10 (low to high) depending on the degree of tremor observed (Table 9.6). The reason we would decide to use a non-parametric test here would be if the differences do not appear to come from a normal distribution. In particular, are there any obvious outliers? If we look at Table 9.7 which gives a difference column we can see all values are generally close together, but there is a clear outlier in the value of −9 for surgeon 7. This justifies our use of a non-parametric test.

In Table 9.7 we have added additional columns to the data from Table 9.6, to assist in explaining how the test works. In order to do the calculations manually, first the differences between coffee and hot water are calculated and then the ranks of those differences are calculated for the absolute values (i.e. irrespective of whether they are positive or negative values). The procedure for dealing with ties is the same as we used in the Mann–Whitney test above.

Table 9.7 Additional columns added to the data given in Table 9.6. As these are paired observations, that is, taken from the same participants, we need to subtract the hand tremor scores (difference). As explained in the text the two additional columns give ranking information

Surgeon	Hot water	Coffee	Difference	Ranked diff	Signed ranks
1	2	8	6	4	4
2	3	4	1	1	1
3	1	7	6	4	4
4	0	9	9	7	7
5	4	10	6	4	4
6	2	6	4	2	2
7	10	1	−9	7	−7
8	1	10	9	7	7

Finally the last column, signed ranks, has the ranks preceded by their sign (– if the difference was negative and unchanged from the previous column if the difference was positive).

We could perform a sign test (beginning of chapter) on the data in the last column, with one negative value (comparing differences from 0), giving $p = .07$, which is not statistically significant. The procedure we will use is variously called Wilcoxon test, Wilcoxon's matched-pairs signed-ranks test, Wilcoxon signed-rank test (easily confused with one of the names for the independent samples procedure above). Proceeding with the ranking method, the totals of the signed ranks are 7 for the negative ranks and 29 for the positive ranks. Either of these can be taken as the V statistic to report.

Again all statistical packages will run the test. The R code and outputs are given in Box 9.3. We can see from the p value that statistical significance is not achieved.

Box 9.3   R code for entering data, obtaining medians and sample sizes, and running the Wilcoxon paired samples test

```
> shakes <- c(2,3,1,0,4,2,10,1,8,4,7,9,10,6,1,10)          #enter data
> treat <- c(rep("Hot water",8), rep("Coffee",8))
> caffeine <- data.frame(shakes,treat)
> tapply(shakes,treat,median)                              #gives medians
  Coffee Hot water
     7.5       2.0
> tapply(shakes,treat,length)                              #gives sample size
  Coffee Hot water
       8         8
> wilcox.test(shakes~treat,paired=TRUE,conf.int=FALSE)     #wilcoxon signed rank test

        Wilcoxon signed rank test with continuity correction

data:  shakes by treat
V = 29, p-value = 0.1376
alternative hypothesis: true location shift is not equal to 0

Warning message:
In wilcox.test.default(x = c(8, 4, 7, 9, 10, 6, 1, 10), y = c(2,  :
  cannot compute exact p-value with ties
```

Results summary: A Wilcoxon signed-rank test showed that there was no significant difference in the amount of hand tremor between the hot water and coffee treatments, hot water median = 2, coffee median = 7.5, $V = 29$, $N = 8$, $p = .138$. A boxplot of the data is given in Figure 9.3.

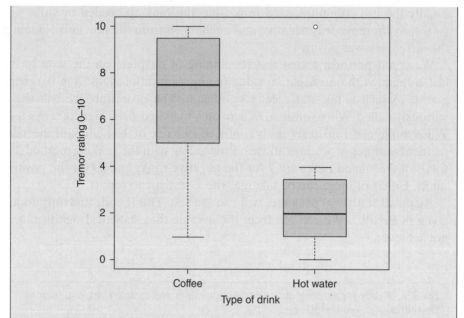

Figure 9.3 Boxplot of the effect of coffee versus placebo on the amount of hand tremor exhibited by surgeons. In the hot water treatment, there is an outlier with a tremor rating of 10, indicated by a circle.

Like the independent samples test, if $N > 20$ then we can calculate a z value. The formula to use is

$$z = \frac{\left(S + \frac{1}{2} - \left(\frac{1}{4} * N * (N + 1)\right)\right)}{\sqrt{[N * (N + 1) * (2 * N + 1) / 24]}}$$

S is the smaller of the summed signed ranks (in this example it would be 7).

If there are zero differences then these need to be summed. We denote this sum as d_0. For example, if there were three zeros, then $d_0 = 3$. In our example, there are no zero differences so $d_0 = 0$. If there are ties in the ranks, each number of ties per group denoted d_i, then these need to be calculated as we did previously above, for example, using our example $d_1{}^3 - d_1 = 24$ and $d_2{}^3 - d_2 = 24$. These are summed in the denominator of the equation below.

$$z = \frac{\left(S - \left(\frac{1}{4} * (N * (N + 1) - d_0 * (d_0 + 1)\right)\right)}{\sqrt{([N * (N + 1) * (2 * N + 1)] - d_0 * (d_0 + 1)) / 24 - \sum [d_i^3 - d_i] / 48)}}$$

Again we compare the absolute value (i.e. ignore the negative sign) with 1.96 to determine whether the result is significant at the 5% level.

A paired *t*-test performed on this example produces $t(7) = 1.93$, $p = .095$. Therefore, not statistically significant either.

9.7 Kruskal–Wallis one-way analysis of variance

This test is an extension of the Mann–Whitney test for independent groups described above. It allows us to analyse three or more groups, although if it is run with just two groups then it reduces back to the Mann–Whitney test.

In 1973, a study investigated the jumping height of the rat flea. This was measured to be 33 cm (about 130 times its own height). A subsequent important study by Cadiergues *et al.* (2000) assessed jumping heights of two additional species of fleas, those inhabiting cats and dogs. The more recent study was carried out using 18 fleas, 9 fleas per species. We provide additional data for the species inhabiting humans. Nine participant fleas were also enrolled, unfortunately one of these participants escaped the apparatus and could not be traced, although it was believed by the researchers that a new host was found among one of them. The data for that flea were incomplete and therefore removed from the analysis. The data are presented in Table 9.8.

Examination of the table indicates that the data are clearly not normally distributed. For each group there are a cluster of low values (1–13), and then a wider distribution of higher values (20–50). This confirms the need to use a non-parametric test. The ranking procedure is done in the same way as we

Table 9.8 The vertical jumping distances in centimetres of nine cat fleas (*Ctenocephalides felis felis*), nine dog fleas (*Ctenocephalides canis*) and eight human fleas (*Pulex irritans*). Each data point consists of the average of three jumps per flea

Cats	Dogs	Humans
35	8	2
10	36	13
2	42	1
25	30	4
30	5	2
6	50	20
4	3	6
48	47	8
20	49	

Table 9.9   Ranks are to the right of each data column. At the bottom of each rank column are the total ranks, $\sum R$. Tied ranks are highlighted in grey

Cats	Rank	Dogs	Rank	Humans	Rank
35	20	8	11.5	2	3
10	13	36	21	13	14
2	3	42	22	1	1
25	17	30	18.5	4	6.5
30	18.5	5	8	2	3
6	9.5	50	26	20	15.5
4	6.5	3	5	6	9.5
48	24	47	23	8	11.5
20	15.5	49	25		
$\sum R_c =$	127	$\sum R_d =$	160	$\sum R_h =$	64

did for the Mann–Whitney test. Again, we need to rank the values irrespective of their group membership: the smallest rank of 1 is given to the human flea height of 1 cm, and the largest rank of 26 is given to the dog flea height of 50 cm (Table 9.9).

The statistic we calculate is known as H:

$$H = \frac{12}{N*(N+1)} * \sum \frac{R_k^2}{n_k} - 3*(N+1)$$

where k is the number of groups, n is the number in each group and N is the total number of fleas in the study.

$$H = \frac{12}{26*(26+1)} * \left(\frac{127^2}{9} + \frac{160^2}{9} + \frac{64^2}{8} \right) - 3*(N+1)$$

$$H = 7.009496676$$

This has to be corrected for ties. The correction factor is calculated by:

$$1 - \frac{\sum T_i}{N^3 - N}$$

Where $T_i = t_i^3 - t_i$

t_i is the number of tied values in the ith group of tied scores. For example, the first tied group comprises three values of two, giving $T = 3^3 - 3 = 27 - 3 = 24$. The remaining five groups consist of pairs of ties, each $T = 2^3 - 2 = 6$.

So that the correction factor is

$$1 - \frac{(24 + 6 + 6 + 6 + 6 + 6)}{26^3 - 26}$$

$$= 0.996923077$$

The *H* value calculated earlier is then divided by this correction factor:

$$H = 7.009496676 \,/\, 0.996923077$$

$$= 7.031130925$$

This value is equivalent to a χ^2 value with $k - 1$ degrees of freedom (here df $= 3 - 1 = 2$), which can be compared with the critical values shown in Table 8.1, or from Excel or R. The R code and output are given in Box 9.4.

Box 9.4   R code for calculating medians, sample sizes, and performing the Kruskal–Wallis one-way analysis of variance

```
> height <- c(35,10,2,25,30,6,4,48,20,8,36,42,30,5,50,3,47,49,2,13,1,
4,2,20,6,8)
> animal <- as.factor(c(rep("Cat",9),rep("Dog",9),rep("Human",8)))
> dat1 <- data.frame(height,animal)
> tapply(height,animal,median)                  #gives medians
  Cat    Dog Human
   20     36     5
> tapply(height,animal,length)                  #gives sample sizes
  Cat    Dog Human
    9      9     8
> kruskal.test(height~animal)                   #Kruskal-Wallis test

        Kruskal-Wallis rank sum test

data:  height by animal
Kruskal-Wallis chi-squared = 7.0311, df = 2, p-value = 0.02973
>
> pchisq(7.031,df=2,lower.tail=FALSE)    #p value for chi-square= 7.031, 2 df
[1] 0.02973293
> qchisq(0.950, 2)             #critical chi-squared 5% signif 2 df
[1] 5.991465
```

Results summary: A Kruskal–Wallis one-way analysis of variance compared the jumping heights of three different species of flea. Two or more species were significantly different from each other, $\chi^2(2) = 7.03$, $p = .030$. Dog fleas jumped the highest (median $= 36$ cm, $n = 9$) > cat fleas (median $= 20$ cm, $n = 9$) > human fleas (median $= 5$ cm, $n = 8$). A boxplot is given in Figure 9.4.

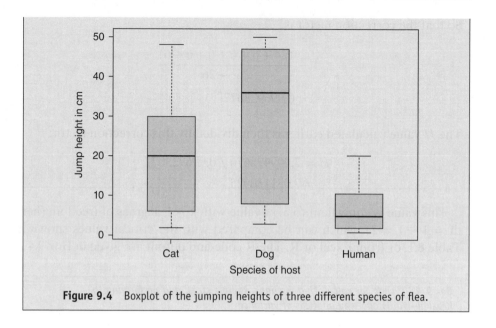

Figure 9.4   Boxplot of the jumping heights of three different species of flea.

With such a low jump height for human fleas, the experimenters now feel fairly safe from catching the escaped flea.

We know from the results of the analysis that two or more flea species differ from each other, but we don't know which ones. It is possible that all three species differ from each other. We can use a Mann–Whitney test pairwise to test for differences, that is, dog fleas versus cat fleas, dog fleas versus human fleas, cat fleas versus human fleas. See if you can do them, and report the results.[2] Since there are only three groups and the overall analysis was statistically significant, it is not necessary to adjust the procedure to guard against Type I errors. However, if there are more than three groups then an adjustment should be made to the significance level using Bonferroni correction, so that $\alpha = .05$/number of comparisons. We did a single overall test (Kruskal–Wallis) to see if there were any differences at all, and if there weren't any, it saved us the time of running lots of multiple comparisons.

A parametric one-way ANOVA produces $F_{2,23} = 4.80, p = .018$, a value that is slightly more statistically significant than that obtained above. However, we decided to use the non-parametric test because the data in each group did not look normally distributed. By using a non-parametric test we were adopting a cautious approach.

---

[2] Only the dog fleas versus human fleas was statistically significant. Dog fleas (median = 36, $N = 9$) jumped significantly higher than human fleas (median = 5, $N = 8$), $U = 11.5, p = .015$. Dog fleas were not significantly different from cat fleas (median = 20, $N = 9$), $U = 26.5, p = .222$. Cat fleas were marginally not significantly different from human fleas, $U = 16.5, p = .059$.

9.8 Friedman test for correlated samples

This test is the non-parametric equivalent for the repeated measures analysis mentioned in Chapter 6. Where we have more than two treatments, and where the scores are related across the different treatments. Typically, the scores are related because the same participants are tested in each of the different treatments. We could imagine that our surgeons and coffee example above had another treatment, decaffeinated coffee (Table 9.10). This would act as a more appropriate placebo than hot water since it would (presumably) taste and look the same as the coffee treatment. The order in which the three treatments were administered would also need to be randomised or counterbalanced to control for order effects.

The ranking is then performed **within each participant's** set of scores, with the highest value for each participant ranked 3 and the lowest ranked 1 (Table 9.11).

Table 9.10   The data from Table 9.6 with a decaffeinated coffee treatment added (third column)

Surgeon	Hot water	Decaf coffee	Coffee
1	2	3	8
2	3	1	4
3	1	3	7
4	0	8	9
5	4	0	10
6	2	0	6
7	10	2	1
8	1	2	10

Table 9.11   Data from Table 9.10 converted to ranks within each participant

Surgeon	Hot water	Decaf coffee	Coffee
1	1	2	3
2	2	1	3
3	1	2	3
4	1	2	3
5	2	1	3
6	2	1	3
7	3	2	1
8	1	2	3
Totals	13	13	22

Friedman's χ^2 can be calculated using the following formula:

$$\chi_F^2 = \left(\frac{12}{N * k * (k+1)} * \sum R_i^2 \right) - (3 * N * (k+1))$$

where N is the sample size, k is the number of treatments and R_i^2 are each of the treatment rank totals squared. So we get

$$\chi_F^2 = \left(\frac{12}{8 * 3 * (3+1)} * (13^2 + 13^2 + 22^2) \right) - (3 * N * (k+1))$$

$$= \left(\frac{12}{96} * 822 \right) - 96$$

$$= 6.75$$

This is equivalent to a χ^2 value and should again be evaluated on $k - 1 = 3 - 1 = 2$ degrees of freedom, either using Table 8.1, or from Excel or R. The R code and outputs are given in Box 9.5.

Box 9.5   R code for entering data, obtaining medians and sample size, and running the Friedman test for related samples

```
> hotwater <- c(2,3,1,0,4,2,10,1)                           #enter data
> decaffeine <- c(3,1,3,8,0,0,2,2)
> coffee <- c(8,4,7,9,10,6,1,10)
> coffee3 <- data.frame(hotwater,decaffeine,coffee)
> sapply(coffee3,median)                                    #gives medians
   hotwater decaffeine       coffee
        2.0        2.0          7.5
> sapply(coffee3,length)
   hotwater decaffeine       coffee    #gives number in each treatment
          8          8            8
> friedman.test(as.matrix(coffee3))    #Friedman test, note as.matrix


        Friedman rank sum test

data:   as.matrix(coffee3)
Friedman chi-squared = 6.75, df = 2, p-value = 0.03422
>
> pchisq(6.75,df=2,lower.tail=FALSE)    #p value for chi-square=6.75
[1] 0.03421812
> qchisq(0.950, 2)                       #critical chi-squared for 2 df
[1] 5.991465
```

Results summary: A Friedman test found a statistically significant difference between two or more drink treatments in the amount of hand tremor exhibited by the surgeons, $\chi^2(2) = 6.75$, $N = 8$, $p = .034$. Figure 9.5 gives a boxplot of the data. The coffee treatment produces the highest ratings of tremor (median = 7.5) compared with hot water (median = 2) and decaffeinated coffee (median = 2).

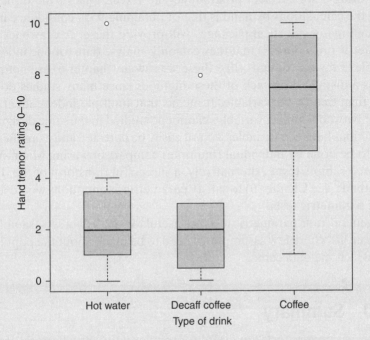

Figure 9.5 Boxplot of the effects of different drinks on the amount of hand tremor in surgeons. The coffee treatment results in the highest tremor ratings (scale 0–10).

A parametric analysis of variance came to a similar conclusion, $F_{2,14} = 4.21$, $p = .037$. This is similar to our value obtained with the non-parametric test. We used the Friedman test here because in the paired samples test we identified an outlier difference. We are again being cautious and also preventing any outliers from interfering with a parametric test (where the variance would be elevated).

For multiple comparisons the non-parametric paired samples test should be used to compare treatments (e.g. coffee versus hot water, coffee versus decaffeinated, hot water versus decaffeinated). If there are more than three treatments, then it will be necessary to correct the significance level using Bonferroni correction. As with the Kruskal–Wallis test, we did a single overall test to see if there were any differences at all, and if there weren't any, it saved us the time of running lots of multiple comparisons.

9.9   Conclusion

All these non-parametric tests can be quite easily done manually, and more easily using a spreadsheet programme like Excel. One of the difficulties in doing the calculations by hand is that of obtaining 95% confidence intervals, although many statistical packages will provide these. A drawback of non-parametric procedures is that they can only analyse data for one independent variable. Two-way designs (like those we saw in Chapter 6) are not possible. This is a distinct drawback of these methods since many studies do involve more than one factor/variable. It means that multiple independent, within- and/or between-subject variables cannot be studied in the same design. Hence interactions between variables cannot easily be detected and examined. These need to be done by individual (multiple) comparisons using Mann–Whitney or paired sample tests. Alternatively, a successful transform of the data (e.g. logarithmic, see Chapter 6) to satisfy parametric assumptions would allow the use of parametric tests.

In general, non-parametric tests are useful for small data sets (sample size of less than 30) where few assumptions need to be made about the population(s) from which the data came.

9.10   Summary

- Non-parametric tests are useful for smaller data sets where we are unable or unwilling to vouch for assumptions required for parametric tests (normality of population, homogeneity of variances). The appropriate central tendency statistic is the median rather than the mean.

- A corresponding non-parametric test is available for each parametric test, except for a multi-way ANOVA, which means that the simultaneous effects of two or more factors cannot be studied in the same analysis. It also means that we are unable to easily detect or examine interactions between the two or more factors.

- The sign test can be done by hand with a calculator or Excel.

- Similarly, the Kruskal–Wallis and Friedman test can be done manually quite easily, both resulting in a χ^2 statistic which can be looked up in a table or the critical value can be obtained from Excel (e.g. $=CHISQ.INV(0.95,1)$ for 2010 version or $=CHIINV(0.05,1)$ for previous versions) or R (e.g. $qchisq(0.950, 1)$). Or better, the exact p value can be obtained using $=CHIDIST(value,df)$, $=CHISQ.DIST.RT(value,df)$, $pchisq(value,df=DF,lower.tail=FALSE)$ in Excel and R respectively.

- It is more problematic doing the Mann–Whitney and Wilcoxon tests manually since they result in less well-known statistics (U and V). With larger samples ($N > 20$ for each treatment), z can be calculated and compared with the critical value 1.96 or larger, depending on the significance level required. It is easier to use a computer package which will give exact p values.

- Computer packages will also calculate 95% confidence intervals which are difficult to do manually.

- When more than two treatments (levels) are included then multiple comparisons can be performed pairwise using the corresponding non-parametric independent samples and paired samples tests. If more than three treatments are present, then a Bonferroni correction to the significance level should be applied.

References

Cadiergues, M-C., Joubert, C. and Franc, M. (2000) A comparison of jump performances of the dog flea, *Ctenocephalides canis* (Curtis, 1826) and the cat flea, *Ctenocephalides felis felis* (Bouché, 1835). *Veterinary Parasitology*, 92:239–241.

Sprent, P. (1992) *Applied Nonparametric Statistical Methods*. Chapman & Hall/CRC.

10
Resampling Statistics comes of Age – 'There's always a Third Way'

10.1 Aims

Although resampling techniques have been around for decades, their popularity was initially restricted by a lack of computing power sufficient to run thousands of iterations of a data set. Nowadays, with more powerful modern computer and the rapid expansion of genomics, the acquisition of extremely large data sets is reasonably common and requires statistics beyond the scope of classical parametric or non-parametric tests. This chapter is intended simply as an introduction to the subject and will explain how resampling statistics are an attractive alternative as they can overcome these limitations by using non-parametric techniques without a loss of statistical power. This chapter will be of particular relevance to biomedical sciences students, and psychology students interested in the field of cognitive neuroscience.

10.2 The age of information

DNA microarrays heralded the era of genomics and previously unimaginably large data sets. Instead of meticulously working on one gene, laboriously investigating the factors regulating its expression, its expression patterns and the relationship of mRNA to protein levels (if any), researchers now investigate thousands of genes at a time. And not only their expression. Next, generation sequencing costs have diminished substantially and are now within the reach of many labs. These techniques have changed the way we perceive the

Starting Out in Statistics: An Introduction for Students of Human Health, Disease, and Psychology
First Edition. Patricia de Winter and Peter M. B. Cahusac.
©2014 John Wiley & Sons, Ltd. Published 2014 by John Wiley & Sons, Ltd.
Companion Website: www.wiley.com/go/deWinter/startingstatistics

genome. We now know that most of it is actually transcribed and it's not 'junk'. Non-coding RNA is active, it does things like regulate coding genes. So suddenly we have vast amounts of data, tens of thousands of numbers to analyse, not just a few dozen. Similarly, cognitive neuroscientists are now able to image activity of the whole brain using techniques such as functional magnetic resonance imaging (fMRI) and positron emission tomography (PET) scanning. That's around 86 billion neurons and as many glial cells (Azevedo *et al.*, 2009). Although cells are not scanned individually using these techniques, the sheer number of cells present gives some idea of the volume and complexity of data that are generated. OK, so we have computers, but we need alternative statistical methods of analysing these data, because the conventional ones are not fit for purpose. The late David Carradine (of Kill Bill fame) is quoted as saying 'There's an alternative. There's always a third way, and it's not a combination of the other two ways. It's a different way'. We refer to resampling as the third way, the new statistics.

Methods that generate large volumes of data pose a challenge for analysis. Techniques for optical data acquisition such as fluorescence intensity measurement (collected as pixel counts) or medical scans (collected as voxels – think of it as a three-dimensional pixel), invariably involve collection of data that are simply background noise. What do we mean by this? Well, it means that you will measure **something** even if the thing you are trying to measure is absent. This is admirably demonstrated by Bennett *et al.* (2010) who performed an fMRI scan on a dead salmon to which they showed photographs of humans, asking it to determine which emotions were being experienced. The processed scan data reported active voxel clusters in the salmon's brain and spinal column. Clearly, a dead salmon does not have any brain activity and is certainly not able to perform a cognitive task – the results were what we call a false positive and arise from inadequate or lack of correction for multiple testing. As space restrictions are not conducive to analysis of such massive amounts of data, here we will use small data sets as examples to demonstrate the principles of resampling methods, which may then be applied to more complex data such as those encountered in the real world.

10.3   Resampling

Resampling statistics is a general term that covers a wide variety of statistical methods, of which we are going to introduce two: *randomisation tests* and *bootstrapping*. These methods are particularly useful for, but not restricted to, large data sets that are not amenable to conventional tests, such as genomics data. Recently, interest has also been aroused in using bootstrapping for the analysis of fMRI data (e.g. Bellec *et al.*, 2010). In this chapter, we will first demonstrate how randomisation tests and bootstrapping work and then we will discuss their application in the above fields.

Randomisation and bootstrap tests are sometimes called distribution-free tests. What this means is that unlike parametric tests, they make no assumptions about the distribution of the population from which the sample is drawn, and not that there is no distribution. There is always an underlying distribution of data, even if we are unable to discover it. This is an advantage when we have data sets consisting of so many samples that it is not practicable to determine whether each is normally distributed or not. Resampling tests simulate chance by taking thousands of samples from the original data but the sampling method differs between randomisation tests and bootstrapping. We will start with a demonstration of how randomisation tests work. These are sometimes called permutation or exact tests. The greatest flexibility in performing resampling statistics is provided by R, indeed these methods are not yet fully available in many commonly used statistical software packages.

10.3.1 Randomisation tests

The hippocampus is a region of the brain that is important for navigation, a form of spatial memory. You will doubtless know people who have a 'good sense of direction' and others who get lost if they deviate from a learned route. I know a hospital consultant who on his way back to the clinic from a meeting pressed a button with a house symbol on his new satnav and was two streets away from his home before he realised the button did not mean 'reset to the home screen'. So we might ask the question, 'Does the hippocampus of London taxi drivers differ in size than those of their fellow citizens?'. This was indeed a question posed by Maguire *et al.* (2000) of University College London, who compared hippocampal size in 16 licensed male London taxi drivers with those of 50 males who did not drive taxis. We selected this example because the paper was awarded an IgNobel[1] prize in 2003. The authors found that some hippocampal regions were larger in taxi drivers and some were smaller. We will simulate one part of the study, the cross-sectional area of the anterior left hippocampus (ALH), which was smaller in taxi drivers, with approximation of the mean and standard deviation in the original data set but with a sample size of six in each group for simplicity (Table 10.1a).

The null hypothesis is that the area of the ALH in the taxi drivers group does not differ from the area of the ALH in non-taxi drivers. If this null hypothesis is correct, then the particular configuration of observed numbers would be no different from random selections taken from all the numbers of both groups, or to put it another way, any subject could have the same size ALH as any other subject. Now we will jumble up the data so that each

---

[1] The IgNobel Prizes are an American parody of the annual Nobel prizes. They 'honor achievements that first make people laugh, and then make them think'.

Table 10.1 Anterior left hippocampus (ALH) area in London taxi drivers (T) and age-matched non-taxi driver males (C), $n = 6$. (a) Original and (b) one permutation of the same data and (c) a second permutation of the same data

(a) Difference in means = 80.8 − 101.3 = −20.5

Subject	Area of ALH (mm²)	Group mean
T1	70	
T2	72	
T3	79	
T4	80	
T5	84	
T6	100	80.8
C1	90	
C2	89	
C3	109	
C4	100	
C5	108	
C6	112	101.3

(b) Difference in means = 86.7 − 95.5 = −8.8

T1	89	
T2	72	
T3	79	
T4	100	
T5	100	
T6	80	86.7
C1	90	
C2	70	
C3	108	
C4	84	
C5	109	
C6	112	95.5

(c) Difference in means = 94.6 − 87.5 = 7.2

T1	112	
T2	100	
T3	79	
T4	108	
T5	89	
T6	80	94.7
C1	72	
C2	100	
C3	90	
C4	70	
C5	109	
C6	84	87.5

value is randomly allocated to a subject irrespective of the group to which he belongs. We end up with a new set of randomised data, or one *permutation*, hence the origin of the test name (Table 10.1b). By chance some subjects will have been allocated exactly their own results (i.e. subjects T2, T3, C1 and C6), some subjects will have been allocated values for another subject within the same group (i.e. T4, T6, C3 and C5) and some subjects have ended up with values for the other group (i.e. T1, T5, C2 and C4). If we calculate the new mean for each group we get 86.7 for the taxi drivers and 95.5 for the non-taxi drivers. We repeat this yet again and end up with yet another permutation of the data with a mean of 94.7 for the taxi drivers and 87.5 for the non-taxi drivers (Table 10.1c). Now imagine that we repeat this randomisation step say, 100,000 times (this is the bit where a computer is needed). By chance, we will end up with many data sets that look nothing like the original and some that will look similar, if not identical.

Next we calculate some property of the data set, which we will call the test statistic – in this example I will use the difference in means – for the original data and for all 100,000 permutations of the data. The test statistic for the original data is $80.8 - 101.3 = -20.5$. Now, if we plot a histogram of all the test statistics (differences in means) that we have calculated, it would look like Figure 10.1. As we have access to the original data, we can tell you that there are 451 test statistics lower (more extreme) than the difference in means of -20.5 of our original sample although it is tricky to determine this from the graph. There are also 424 test statistics with a difference in means greater than 20.5. This is $424 + 451 = 875$ test statistics that are more extreme than the difference in means between the original samples. To calculate the probability

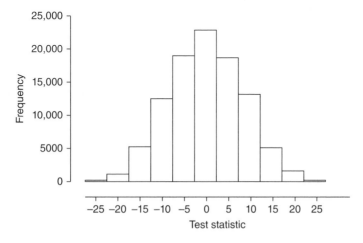

Figure 10.1 Histogram of the test statistic 'difference in means' for 100,000 permutations of the data in Table 10.1.

for a two-tailed test, divide this total number of test statistics by the number of permutations: 875 / 100,000 = .00875, which is significant at the 1% level.

Note that because the permutations of the original data are random, if we were to repeat the test again, we might get a slightly different result. This differs from a *t*-test, which if repeated on the same set of data will yield exactly the same result each time. However, as we have run a large number of permutations for the randomisation test, we are likely to achieve a very similar result rather than an identical one. So the probability obtained by a randomisation test is an approximation. Another point to note is that unlike a *t*-test where we calculate a confidence interval (usually 95%) to estimate the possible locations of the population mean, we have made no estimate of any population parameters with the randomisation test. Randomisation tests cannot estimate population parameters. All they tell us is how many times the property of our data that we have selected as the test statistic is exceeded by chance alone, by resampling the original numbers as random permutations. We used a difference in means as the test statistic for our data, but we could have used a different measure, for example, the value of *t*. For randomisation tests, one test statistic is as good as another if it gives a similar result. We used the difference in means here because it is easier to calculate than *t*. Our conclusion from the randomisation test that we have just performed is that the left anterior hippocampi of taxi drivers is significantly smaller in cross-sectional area than that of non-taxi driver males by a mean value of -20.5 mm^2, $p < .01$.

10.3.2   Bootstrapping

Bootstrapping is a method that also comes under the general umbrella term of resampling statistics. Bootstrapping will provide slightly different results each time it is repeated but is approximate enough to be practically useful even though it does not provide the exact *p* values generated by permutation tests. The strength of bootstrapping is that it does not make any assumptions about the properties of the population from which the sample was taken, the only requirement is that it is a random sample. So with no information about the population the best information we have about the population is the values in our random sample. In previous chapters we have used the sampling distribution of a sample statistic (such as *t*, or *r*) to test hypotheses, and this sampling distribution is composed of many random samples from the population. By repeated sampling from our single random sample, the bootstrap produces a distribution that closely mimics the sampling distribution, including pretty accurate estimates (little bias) of the *sample* statistics, such as mean or standard deviation. When a sample is from a normal population the parameters of the sampling distribution, sample mean and standard deviation are pretty good estimates of *population parameters* (little bias) and can be used in hypothesis testing. The usefulness of the bootstrap is that as the bootstrap distribution mimics the sampling distribution, then (a) we can use the bootstrap distribution to test for normality of the sampling distribution and (b) if the bias of sample statistics of the bootstrap distribution and bias of these for the sampling distribution are pretty close *then as a bootstrap distribution mimics the sampling distribution* we can use the bootstrap for testing hypotheses about the population. In the examples here we examine the use of the bootstrap in setting confidence intervals, testing for normality and comparing two groups. Lastly, we consider briefly if normality of the bootstrap distribution is not the case using the example of the correlation coefficient. However, none of this would be possible without the availability of cheap and powerful computers as the numbers of computations are considerable.

First, we are going to demonstrate the bootstrap method of sampling with *replacement* using some data for systolic blood pressure during rest and during exercise in undergraduate students. The raw data are provided in Appendix A. Unlike randomisation tests where we scramble the original data for every subsequent sampling, the bootstrap samples with replacement. This means that if we have 49 numbers in the original sample, in order to construct the first bootstrap sample, the computer will select a value from the original but replace it before it selects the next number, so that first number might be drawn again in any given bootstrap sample. It is analogous to having a bag containing numbered balls, pulling out a ball, writing down its number and putting it back in the bag before drawing out a second ball, which could by chance then be the same one again. Consequently, in any bootstrap sample, some numbers will by chance appear more times than they do in the original data, some fewer times, some the same number of times and some not at all (Table 10.2, a colour version is located in the the colour plate section).

Our original sample exhibits positive skewness, but as this is a bootstrap test (Figure 10.2) what is important is the shape of the bootstrap distribution. Now we'll take 10,000 bootstrap samples from the original random sample and for each we calculate the mean: the bootstrap 1 mean is 113.9, the bootstrap sample 2 mean is 111.7, etc. Now we can plot a histogram of all 10,000 bootstrap means (Figure 10.3a). This distribution of bootstrap means is pretty symmetrical and the Q–Q plot indicates little deviation from normality (Figure 10.3b), so we can also ask the software to calculate a 95% confidence interval: if the means are ranked lowest to highest the lower 95% confidence limit is the mean for the 250th bootstrap sample (250 is 2.5% of 10,000) and the upper 95% confidence limit is the mean for the 9750th bootstrap sample (9750 is 10,000 − 250). This is called the *percentile* method and for our data the 95% confidence interval is 111.5918–116.5306 mmHg. The R commands used to generate the histogram and the percentile method 95% confidence

Table 10.2 Demonstration of the bootstrap resampling method

Original	Bootstrap 1	Bootstrap 2	Original	Bootstrap 1	Bootstrap 2
99	99	99	112	113	110
103	103	99	113	113	112
104	104	103	113	113	112
104	104	103	113	113	113
104	106	103	113	113	113
105	106	104	113	113	113
106	106	104	113	113	113
107	107	104	113	113	113
107	108	105	114	113	113
108	108	106	115	113	113
108	108	106	115	113	114
108	108	107	115	113	114
109	109	107	117	115	115
109	109	108	117	115	115
110	110	108	118	118	115
110	111	108	118	118	115
110	111	108	120	118	118
111	111	108	126	128	118
111	112	108	128	128	118
111	112	109	130	130	126
112	112	109	130	130	126
112	112	109	135	130	128
112	112	109	135	130	130
112	112	109	142	135	135
112	113	110			

The bootstrap samples were randomly resampled with replacement from the column of original data, which are resting systolic blood pressures ranked low to high $n = 49$, $\bar{x} = 113.9$ mmHg. Values in blue represent numbers sampled the same number of times in the bootstrap samples as they occurred in the original data. Values in green appear more times in the bootstrap sample than the original, for example, there is one value of 106 in the original sample but it has been sampled three times in bootstrap sample 1 and twice in bootstrap 2. Values in orange appear fewer times in the bootstrap samples than in the original, that is, 110 appears three times in the original sample but only once in bootstrap sample 1 and twice in bootstrap sample 2. Finally, values in the original samples may be omitted from a bootstrap sample, for example, 105 does not appear in bootstrap 1 and 111 is also missing in bootstrap 2. (For a colour version, see the colour plate section.)

intervals are reported in Box 10.1. Other parameters may be estimated from the distribution of bootstrap means, for example, the standard error.

Now let's be clear about what this 95% confidence interval tells us. It is not necessarily an estimate of where the population mean lies, but rather of where the mean of the original sample lies for repeated bootstrapping samples. But if the distribution of the bootstrap means is symmetrical and normal, and the bias is small then the confidence interval may be used to estimate the location of the bootstrap mean (or any sample statistic), but where bootstrap distribution is skewed it tends to be less accurate (biased). The bias in this example is

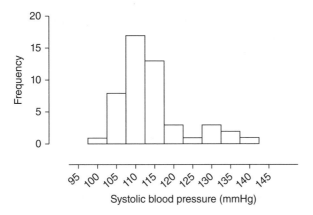

Figure 10.2 Histogram of original sample of 49 resting systolic blood pressure measurements. The sample distribution is positively skewed.

quite small (0.024, Box 10.1) compared with the sample mean value, and the histogram and Q–Q plot indicate little deviation of the bootstrap sample distribution from normal (Figure 10.3). Entire chapters in books on resampling statistics are devoted to the discussion of the efficacy of different confidence interval methods to correct for this bias, but these methods are beyond the scope of this book. However, the confidence interval command in R does give you the estimates of 95% CI for some of the other, less biased methods.

While our histogram of bootstrap sample means is reminiscent of the central limit theorem, it is quite not the same thing. The central limit theorem

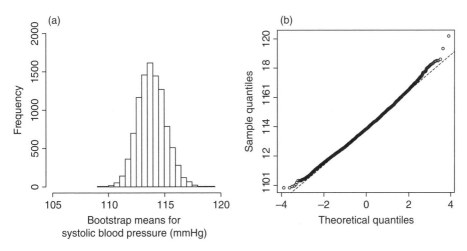

Figure 10.3 (a) Histogram and (b) normal Q–Q plot generated in R for 10,000 bootstrap samples from an original sample of 49 resting systolic blood pressures in Table 10.2.

Box 10.1   *R commands used to generate 10,000 bootstrap samples of 49 resting systolic blood pressure measurements and calculate the percentile 95% confidence interval. Comments are in uppercase*

USING R SAMPLE FUNCTION THE ORIGINAL SAMPLE IS RESAMPLED (WITH REPLACEMENT, REPLACE=TRUE) AND MEAN COMPUTED FOR EACH AND PLACED IN VECTOR SYSR.BOOT. THESE VALUES ARE PLOTTED IN FIGURE 10.3

```
> systolic <- read.csv("Systolic.csv")
> attach(systolic)
> set.seed(5)    #to replicate example here
> sysr.boot<-numeric(10000)
> for (i in 1:10000){sysr.boot[i]<-mean(sample(systr,replace=TRUE))}
```

THE MEAN OF THESE RESAMPLE MEANS AND 95% CONFIDENCE LIMITS ARE COMPUTED BY

```
> mean(sysr.boot)
   [1] 113.9177
> quantile(sysr.boot,c(0.025,0.975))
     2.5%        97.5%
111.5918 116.5306
```

R HAS A BUILT-IN BOOTSTRAP FUNCTION WHICH COMPUTES THE BOOTSTRAP MEAN AND THE BIAS, THE EXTENT TO WHICH THE BOOTSTRAP MEAN DIFFERS FROM THE ORIGINAL SAMPLE MEAN.

```
> install.packages("boot")
> library(boot)
> sysr.stat<-function(systr,i) mean(systr[i])
> sysr.rboot<-boot(systr,sysr.stat,R=10000)
> sysr.rboot
```

```
ORDINARY NONPARAMETRIC BOOTSTRAP
Call: boot(data = systr, statistic = sysr.stat, R = 10000)


Bootstrap Statistics :
    original       bias        std. error
t1* 113.9184    -0.01401633    1.243739
```

AND ON THE RESAMPLE DATA THIS COMMAND CALCULATES THE DIFFERING FLAVOURS OF CONFIDENCE INTERVALS, FOR 4 DIFFERENT METHODS INCLUDING THE SIMPLE PERCENTILE METHOD COMPUTED ABOVE.

```
> boot.ci(sysr.rboot)
```

```
BOOTSTRAP CONFIDENCE INTERVAL CALCULATIONS
Based on 10000 bootstrap replicates


CALL : boot.ci(boot.out = sysr.rboot)
```

```
Intervals :
Level        Normal              Basic
95%     (111.5, 116.4 )    (111.4, 116.3 )


Level        Percentile          BCa
95%     (111.6, 116.4 )    (111.8, 116.7 )


> detach(systolic)


INCIDENTALLY, IF THE CODE IS RUN OMITTING THE set.seed(5) COMMAND THEN THE RESAMPLING
WILL PRODUCE A SLIGHTLY DIFFERENT OUTCOME, WHICH ILLUSTRATES NICELY THE RANDOM ELEMENT
INVOLVED IN BOOTSTRAPPING.
```

is concerned with the distribution of means of many random samples taken from the **population**. In bootstrapping we are repeatedly resampling from **one** sample of the population, and we are using these bootstrap samples to model the sampling distribution. But a similarity between sampling distribution and bootstrap distribution explained above does mean that given normality for the bootstrap distribution we can use the bootstrap to test hypotheses comparing two groups.

10.3.3   Comparing two groups

One application of the bootstrap is to compare two groups from each of which we have a random sample. So if we were to compare our resting systolic blood pressure sample with one comprising systolic blood pressures measured during exercise we would resample with replacement 10,000 times and calculate means for each of the two group samples as we did before. The bootstrap distributions of means are reported in Figures 10.4a and 10.4b. Note two things, the resting blood pressure bootstrap means are all below 120 mmHg whereas those during exercise are all above 120 mmHg. To compare the two groups, each time a bootstrap sample is drawn from the two original samples, a mean difference was calculated (see Box 10.2 for R commands). This is similar to the randomisation test procedure demonstrated by Table 10.1, it differs only in the method of resampling (since it is done with replacement). In Figure 10.4c the bootstrap distribution of the differences between means are plotted on a histogram and the Q–Q plot (Figure 10.4d) indicate a normal distribution for this bootstrap distribution. Consequently, the bootstrap can be used to test the null hypothesis of no difference in means for the two groups. As the 95% confidence interval for the difference in bootstrap means is −18.921824 to −9.788512 and does not include zero (mean difference of −14.31563) we

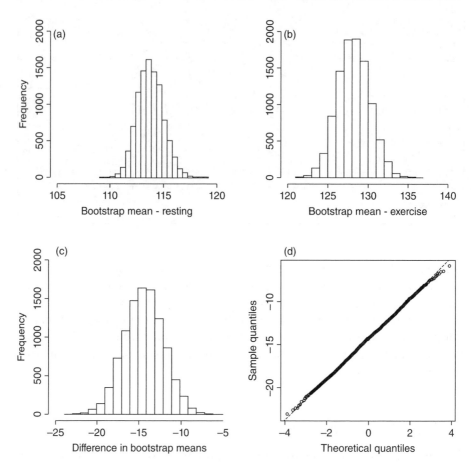

Figure 10.4 Histograms generated in R of 10,000 bootstrap distributions of sample means from (a) an original sample of 49 resting systolic blood pressures (b) an original sample of 31 systolic blood pressures during exercise and (c) the difference in means (b − a) for these bootstrap samples. (d) The normal Q–Q plot for the difference.

would conclude that systolic blood pressure during exercise is higher than that during rest.

Had we wished to compare our two original samples using a parametric method, that is, a *t*-test, we would have had to first transform the data by taking the log of the values as they were not normally distributed (e.g. Figure 10.2). If the transformed data were also not normally distributed, we could have used a non-parametric alternative with the original values, that is, Mann–Whitney *U* test but at the risk of a loss of power – such tests may fail to detect a difference when there is one. The bootstrap test has permitted us to compare the two samples using the original data but without the loss of power sometimes associated with conventional non-parametric tests.

Box 10.2   R commands used to generate 10,000 bootstrap samples of the difference in systolic blood pressure for two independent random samples of resting (n = 49) and exercising (n = 31) undergraduate students and to calculate the percentile 95% confidence interval. Comments are in uppercase

USING R SAMPLE FUNCTION, THE ORIGINAL SAMPLES FOR REST AND EXERCISE ARE RESAMPLED (WITH REPLACEMENT, REPLACE=TRUE) AND MEAN DIFFERENCE COMPUTED FOR EACH AND PLACED IN VECTOR SYSDIFF. THESE VALUES ARE PLOTTED IN FIGURE 10.4.

```
> systolic <- read.csv("Systolic.csv")
> attach(systolic)
> syste <- syste[!is.na(syste)] #this is to remove NA (blanks) from
vector
> set.seed(5)       #to replicate example here
> sysdiff<-numeric(10000)
> for (i in 1:10000){sysdiff[i]<-mean(sample(systr,replace=T))-mean
(sample(syste,replace=T))}
```

MEAN IS COMPUTED

```
> mean(sysdiff)
[1] -14.31563
```

PERCENTILE 95% CONFIDENCE LIMITS

```
> quantile(sysdiff,c(0.025,0.975))
     2.5%       97.5%
-18.921824   -9.78851
```

```
> detach(systolic)
```

10.4   An introduction to controlling the false discovery rate

Thus far we have used some relatively small sample sizes to demonstrate how resampling tests are carried out. So how do we apply these to very large data sets such as genomic or imaging data? Randomisation or bootstrap tests can be applied to many samples at once, all will be resampled in the same way as two are in the examples above and we'll then know which samples are statistically different from each other. The difficulty with genomics and other huge data sets is their multidimensionality; this makes them more complex to analyse. Imagine we investigate the expression of 23,000 genes in each of our samples. Let us suppose that 10,000 genes are actually expressed, we can ignore the 13,000 that are not but we still have an awfully large number of tests to perform. A p value tells us the probability of observing a particular result when the null hypothesis is correct. Therefore, if we performed 10,000 tests, 5% (e.g. five hundred) would be declared statistically significant even if the null hypothesis were true for all of them. So we need to weed out these false positives. We call this controlling the false discovery rate.

Before we go further, let us remind ourselves of statistical error. Type I error is claiming that the null hypothesis is incorrect when it is not and conversely Type II error is claiming that the null hypothesis is correct when it is not (see Chapter 5). It is important to understand when controlling the false discovery rate, that reducing Type II error necessarily increases Type I error and *vice versa*. Hence, reducing the number of false positives needs to be offset against the risk of consequently not detecting targets that are truly different between treatment groups. This balance between the two evils of statistical error may depend upon the application of the data. Consider, for example, a situation where several biomarkers are used diagnostically to distinguish tumour samples from normal tissue samples. In this case, it would be disastrous for the patient if a tumour went undetected, therefore reducing Type II error at the expense of Type I error is probably preferable.

A statistical method that has gained much support in controlling the *false discovery rate* is that of Benjamini and Hochberg (1995). Prior to this, corrections for multiple testing, such as Bonferroni correction, concentrated on the number false positives for **all** tests performed. Bonferroni correction as we have seen in an earlier chapter is simply α, usually .05, divided by the number of tests performed. It doesn't take a great deal of imagination to realise that this is far too stringent; if you have just 5000 tests, threshold p would have to be $\leq.00001$ to find a significant difference in your list of targets. The false discovery rate (FDR) is a less conservative alternative because it corrects for false positives only among tests that are **declared significant**. So we would not use the data for all 10,000 expressed genes, but only those that we decided were differentially expressed. Here, we'll simulate the procedure using a small number of the 200 genes out of the 10,000 that were significantly different ($p < .05$) using multiple randomisation tests.

In the FDR procedure by Benjamini and Hochberg, the threshold p for each test is calculated as follows: first rank all the p values obtained from the significance test from lowest to highest, then starting at the lowest p value, divide its rank (which will be 1 for the lowest) by the total number of tests and multiply by 0.05 (Table 10.3). Compare the result with the original p value. If the calculated value is larger than the original p value, then there is a significant difference for that target, and if the calculated value is smaller than the original p value, then there is no significant difference. Repeat for p value ranked two, and so forth.

The FDR by Benjamini and Hochberg was a landmark innovation in the field of multiple testing and led the way for many further refinements of the

Table 10.3 Example of procedure by Benjamini and Hochberg for controlling the false discovery rate of an example data set of 200 gene targets in total

Original p value	Rank	$\dfrac{Rank * 0.05}{200}$	Significant	Significant using Bonferroni
0.00003	1	0.00025	Y	Y
0.00008	2	0.00050	Y	Y
0.0001	3	0.00075	Y	Y
0.0007	4	0.00100	Y	N
0.0009	5	0.00125	Y	N
0.02	6	0.00150	N	N

For simplicity, the six smallest p values are shown (see text for explanation).

procedure and there are now dozens of variations of their method. FDR procedures may be carried out in R, as well as other software packages used to analyse large multidimensional data sets. Nowadays these types of analyses are carried out by specialist bioinformaticians, so large is the volume of data (terabytes) produced by modern techniques such as next generation sequencing.

10.5 Summary

- Resampling techniques have gained popularity since increased computing power has been available to carry out the large number of permutations of data that these tests employ

- These tests are distribution-free, that is, they do not assume a particular underlying distribution in the data. Furthermore, the power of these tests is generally greater than classical non-parametric tests

- Randomisation tests involve shuffling the data with each permutation, that is, sampling without replacement. Bootstrapping, on the other hand, involves sampling with replacement

References

Azevedo, F.A., Carvalho, L.R., Grinberg, L.T., Farfel, J.M., Ferretti, R.E., Leite, R.E., Jacob Filho, W., Lent, R. and Herculano-Houzel, S. (2009) Equal numbers of neuronal and nonneuronal cells make the human brain an isometrically scaled-up primate brain. *Journal of Comparative Neurology*, 513(5):532–541.

Bellec, P., Rosa-Neto, P., Lyttelton, O.C., Benali, H. and Evans, A.C. (2010) Multi-level bootstrap analysis of stable clusters in resting-state fMRI. *Neuroimage*, 51(3):1126–1139.

Benjamini, Y. and Hochberg, Y. (1995) Controlling the false discovery rate: a practical and powerful approach to multiple testing. *Journal of the Royal Statistical Society. Series B (Methodological)*, 57(1):289–300.

Bennett, C.M., Baird, A.A., Miller, M.B. and Wolford, G.L. (2010) Neural correlates of interspecies perspective taking in the post-mortem Atlantic salmon: an argument for proper multiple comparisons correction. *Journal of Serendipitous and Unexpected Results*, 1(1):1–5.

Maguire, E.A., Gadian, D.G., Johnsrude, I.S., Good, C.D., Ashburner, J., Frackowiak, R.S. and Frith, C.D. (2000) Navigation-related structural changes in the hippocampi of taxi drivers. *Proceeedings of the National Academy of Sciences*, 97(8):4398–4403.

Appendix A: Data Used for Statistical Analyses (Chapters 6,7 and 10)

Chapter 6
Independent *t*-test

Heights of 109 male students						Heights of 122 female students					
59	68	69	71	72	74	59	62	64	65	66	68
62	68	70	71	72	75	59	63	64	65	66	68
62	68	70	71	72	75	60	63	64	65	66	68
64	68	70	71	72	75	60	63	64	65	66	68
65	68	70	71	72		60	63	64	65	66	68
65	69	70	71	73		61	63	64	65	66	68
65	69	70	71	73		61	63	64	65	66	69
65	69	70	71	73		61	63	64	65	66	69
66	69	70	71	73		61	63	64	65	66	69
66	69	70	72	73		61	63	64	65	66	69
67	69	70	72	73		61	63	64	65	66	69
67	69	70	72	73		61	63	64	65	66	69
67	69	70	72	73		62	63	64	65	67	70
67	69	70	72	74		62	63	64	65	67	70
67	69	70	72	74		62	63	64	65	67	70
67	69	70	72	74		62	63	64	65	67	71
68	69	70	72	74		62	63	64	65	67	72
68	69	70	72	74		62	64	64	66	67	
68	69	71	72	74		62	64	65	66	67	
68	69	71	72	74		62	64	65	66	67	
68	69	71	72	74		62	64	65	66	67	

Starting Out in Statistics: An Introduction for Students of Human Health, Disease, and Psychology
First Edition. Patricia de Winter and Peter M. B. Cahusac.
© 2014 John Wiley & Sons, Ltd. Published 2014 by John Wiley & Sons, Ltd.
Companion Website: www.wiley.com/go/deWinter/startingstatistics

One-way ANOVA

Systolic blood pressure	Treatment code
139	1
139	1
144	1
149	1
144	1
143	1
150	1
140	1
142	2
133	2
143	2
139	2
123	2
138	2
138	2
134	2
151	3
145	3
144	3
150	3
145	3
146	3
156	3
145	3
123	4
136	4
125	4
123	4
124	4
130	4
134	4
124	4

	Treatment code
No pills	1
White pills	2
Red pills	3
Blue pills	4

One-way nested ANOVA

Superoxide activity	Cage number	Age code
63.27	1	1
49.44	1	1
54.05	1	1
61.26	2	1
26.48	2	1
51.73	2	1
73.94	3	1
61.87	3	1
58.04	3	1
52.66	4	1
53.93	4	1
51.53	4	1
67.91	1	2
59.31	1	2
62.93	1	2
66.80	2	2
50.79	2	2
64.06	2	2
86.89	3	2
61.92	3	2
60.92	3	2
56.00	4	2
60.78	4	2
47.74	4	2
79.05	1	3
57.60	1	3
63.36	1	3
76.64	2	3
73.87	2	3
63.42	2	3
63.33	3	3
41.30	3	3
63.24	3	3
62.01	4	3
66.89	4	3
64.23	4	3
59.62	1	4
58.25	1	4
65.50	1	4
76.09	2	4
57.83	2	4
66.31	2	4
61.01	3	4
81.26	3	4
68.85	3	4
61.32	4	4
66.50	4	4
85.25	4	4

	Age code
4 weeks	1
10 weeks	2
26 weeks	3
78 weeks	4

Two-way randomised block ANOVA

Migration distance	Treatment code	Block code
200	1	1
230	1	2
250	1	3
550	1	4
600	1	5
350	2	1
350	2	2
390	2	3
500	2	4
620	2	5
430	3	1
480	3	2
410	3	3
550	3	4
590	3	5
550	4	1
560	4	2
490	4	3
630	4	4
690	4	5

	Treatment code
Culture medium	1
Compound a	2
Compound b	3
Compound c	4

	Block code
Night 1	1
Night 2	2
Night 3	3
Night 4	4
Night 5	5

Two-way ANOVA

Systolic blood pressure	Treatment code	Duration code
139	1	2
139	1	2
144	1	2
149	1	2
144	1	4
143	1	4
150	1	4
140	1	4
142	2	2
133	2	2
143	2	2
139	2	2
123	2	4
138	2	4
138	2	4
134	2	4
151	3	2
145	3	2
144	3	2
150	3	2
145	3	4
146	3	4
156	3	4
145	3	4
123	4	2
136	4	2
125	4	2
123	4	2
124	4	4
130	4	4
134	4	4
124	4	4

	Treatment code
No pills	1
White pills	2
Red pills	3
Blue pills	4

	Duration code
Two weeks	2
Four weeks	4

Chapter 7
Linear regression (height and weight for 360 female swimmers – continued on next page)

Height	Weight	Height	Weight	Height	Weight	Height	Weight	Height	Weight
156	45	164	62	168	63	170	50	171	59
156	46	164	58	168	65	170	57	171	72
156	53	165	57	168	59	170	58	172	58
158	49	165	58	168	58	170	61	172	56
158	48	165	56	168	56	170	66	172	58
158	52	165	54	168	55	170	63	172	57
158	56	165	62	168	58	170	54	172	64
159	57	165	65	168	70	170	69	172	55
159	55	165	52	168	54	170	63	172	67
159	53	165	56	168	59	170	67	172	60
160	52	165	56	168	60	170	61	172	62
160	55	165	60	168	68	170	62	172	56
160	59	166	62	168	72	170	66	172	61
160	55	166	54	168	73	170	58	172	62
160	61	166	68	168	66	170	59	173	66
161	57	166	55	168	67	170	61	173	62
161	53	166	53	168	51	170	62	173	67
161	56	166	54	168	52	170	63	173	66
161	55	166	57	168	59	170	62	173	65
161	48	166	50	168	66	170	53	173	65
162	54	166	55	168	58	170	56	173	64
162	55	166	57	168	57	170	58	173	61
162	55	166	56	168	50	170	60	173	64
162	57	166	54	168	63	170	60	173	59
162	53	166	60	168	60	170	64	173	71
162	48	167	64	168	73	170	60	173	68
162	46	167	63	168	56	170	60	173	67
162	54	167	55	168	57	170	62	173	69
162	56	167	62	169	58	170	61	174	66
162	52	167	67	169	56	170	62	174	65
163	59	167	58	169	60	170	62	174	60
163	61	167	63	169	58	170	63	174	71
163	55	167	74	169	58	171	56	174	68
163	57	167	68	169	61	171	61	174	66
163	54	167	52	169	59	171	67	174	65
163	54	167	53	169	55	171	66	174	67
163	53	167	62	169	59	171	62	174	69
164	56	167	56	169	59	171	62	174	76
164	58	167	58	169	57	171	61	174	65
164	56	167	64	169	58	171	60	174	60
164	59	168	62	169	53	171	59	174	63
164	58	168	62	170	55	171	59	174	67

Linear regression data continued

Height	Weight	Height	Weight	Height	Weight	Height	Weight
174	73	177	67	179	65	186	72
174	62	177	62	179	66	186	70
174	65	177	67	180	68	186	71
174	66	177	70	180	67	186	74
174	55	177	69	180	67	186	76
175	57	177	69	180	69	187	73
175	68	177	65	180	67	187	73
175	65	177	69	180	71	187	78
175	63	177	63	180	68	188	82
175	61	177	58	180	66	188	81
175	62	177	67	180	69	188	79
175	65	177	68	180	65	188	73
175	61	177	67	180	68	189	74
175	65	177	64	180	69	189	79
175	69	177	66	180	69	189	87
175	62	177	75	180	72		
175	62	177	64	180	66		
175	61	177	60	180	63		
175	64	177	71	181	68		
175	61	178	76	181	69		
175	63	178	67	181	68		
176	64	178	63	181	74		
176	62	178	62	181	67		
176	58	178	60	181	64		
176	61	178	70	181	68		
176	62	178	67	181	70		
176	71	178	66	182	72		
176	66	178	69	182	69		
176	65	178	62	182	73		
176	68	178	66	182	71		
176	63	178	70	182	71		
176	65	178	58	183	73		
176	63	179	61	183	70		
176	60	179	58	183	72		
176	65	179	68	183	74		
176	64	179	65	184	70		
176	63	179	67	184	79		
176	64	179	70	184	78		
176	63	179	66	184	71		
176	70	179	73	184	71		
176	74	179	64	185	76		
177	68	179	62	185	75		
177	68	179	64	185	73		
177	66	179	67	185	72		
177	66	179	69	185	78		

Correlation (maternal BMI and infant birth weight for 100 women and their babies)

BMI	Birth weight	BMI	Birth weight	BMI	Birth weight
17.0	3535	23.1	3648	29.0	3745
16.0	3406	23.2	3562	29.3	3780
15.0	3389	23.4	3665	29.0	3845
18.5	3513	23.5	3772	33.4	3915
18.5	3555	23.5	3623	34.0	3889
18.6	3440	23.5	3707	34.5	3866
18.6	3503	23.5	3739	31.0	3900
18.7	3450	23.8	3623	32.0	3866
19.0	3560	23.9	3598	33.0	3785
19.0	3512	24.0	3760	31.0	3765
19.0	3588	24.0	3691	37.0	3860
19.5	3445	24.0	3722	40.0	3906
19.5	3650	24.0	3664		
19.6	3568	24.0	3749		
20.0	3523	24.2	3780		
20.0	3613	24.4	3728		
20.0	3554	24.5	3588		
20.5	3598	24.5	3654		
20.5	3631	24.7	3785		
21.0	3650	24.8	3686		
21.0	3510	24.9	3764		
21.0	3554	24.9	3782		
21.1	3608	24.9	3624		
21.4	3667	24.9	3715		
21.4	3580	20.6	3546		
21.5	3644	25.3	3669		
21.5	3701	27.0	3740		
21.9	3537	27.5	3730		
22.0	3608	25.7	3722		
22.0	3746	25.4	3780		
22.0	3632	26.2	3695		
22.0	3688	29.0	3712		
22.0	3717	28.5	3713		
22.2	3588	29.6	3730		
22.3	3679	27.0	3612		
22.4	3644	27.5	3852		
22.5	3715	28.9	3825		
22.5	3610	28.4	3851		
22.8	3652	28.5	3675		
22.8	3695	28.0	3771		
23.0	3690	29.9	3805		
23.0	3750	26.1	3754		
23.0	3786	27.6	3780		
23.0	3485	26.7	3791		

Chapter 10
Bootstrapping

Systolic blood pressure at rest		Systolic blood pressure during exercise
99	120	112
103	126	112
104	128	115
104	130	116
104	130	117
105	135	118
106	135	119
107	142	120
107		120
108		120
108		120
108		122
109		122
109		122
110		125
110		126
110		126
111		127
111		129
111		131
112		132
112		135
112		136
112		137
112		141
112		144
113		145
113		145
113		145
113		147
113		149
113		
113		
114		
115		
115		
115		
117		
117		
118		
118		

Appendix B: Statistical Software Outputs (Chapters 6–9)

Chapter 6

For R: Lines starting with '>' are command lines with code, while text after '#' are comments. The remaining lines are outputs.

R for independent two-sample *t*-test

Read in the variables in two columns (and if sample sizes are unequal the missing values should be substituted with NA). If variances known, or assumed equal, then the output returns a value for *t*, the *p* value. Additionally, the output contains the 95% confidence interval of the mean difference and sample means.

```
> height  <-  read.csv("MaleFemaleHeights.csv")
> attach(height)
> t.test(male,female,var.equal=TRUE)


        Two Sample t-test
data:  male and female
t = 14.9129, df = 229, p-value < 2.2e-16
alternative hypothesis: true difference in means is not equal to 0
95 percent confidence interval:
 4.628164  6.037350
sample estimates:
```

Starting Out in Statistics: An Introduction for Students of Human Health, Disease, and Psychology
First Edition. Patricia de Winter and Peter M. B. Cahusac.
© 2014 John Wiley & Sons, Ltd. Published 2014 by John Wiley & Sons, Ltd.
Companion Website: www.wiley.com/go/deWinter/startingstatistics

```
mean of x    mean of y
70.04587     64.71311
```

The two-tailed critical value of t for these degrees of freedom can be found by the $qt()$ command. The value of t here exceeds this value and so we would reject H_0 at $p < 0.05$, or quote the exact p value, which here is very small.

```
> (0.975,229)
[1] 1.970377
```

If variances assumed, or known to be unequal, then the degrees of freedom are adjusted to produce a more conservative test.

```
> t.test(male,female)
        Welch Two Sample t-test
data:   male and female
t = 14.8099, df = 217.031, p-value < 2.2e-16
alternative hypothesis: true difference in means is not equal to 0
95 percent confidence interval:
 4.623054    6.042459

sample estimates:
mean of x    mean of y
 70.04587     64.71311


> detach(height)    #finish analysis by detaching object
```

Using SPSS – independent two-sample *t*-test

For *Analyse/Compare Means/Independent Samples t-Test*, the variable should be in a single column with the grouping codes in the adjacent column. They should be entered in the *Test variable* box and the grouping variable, gender in this case, in the *Grouping Variable* box. The *Define groups* button allows the codes to be defined (here, male – 1, female – 2).

The following commands produce the following output.

```
T-TEST GROUPS=gender(1 2)
  /MISSING=ANALYSIS
  /VARIABLES=height
  /CRITERIA=CI(.95).
```

	Group Statistics				
	Gender	N	Mean	Std. Deviation	Std. Error Mean
Height	1	109	70.05	2.885	.276
	2	122	64.71	2.550	.231

		Levene's Test for Equality of Variances	
		F	Sig.
Height	Equal variances assumed	.643	.423
	Equal variances not assumed		

		t-test for Equality of Means						
							95% Confidence Interval of the Difference	
		T	df	Sig. (2-tailed)	Mean Difference	Std. Error Difference	Lower	Upper
height	Equal variances assumed	14.913	229	.000	5.333	.358	4.628	6.037
	Equal variances not assumed	14.810	217.031	.000	5.333	.360	4.623	6.042

SPSS produces a voluminous output, which includes Levene's test for equality of variances (if Sig. $p < .05$ then variances are not equal) and the statistics for both equal variance and unequal variance t-tests. In addition, the output also lists the mean difference, its standard error, and the confidence interval of the difference. Q-Q plots are readily obtained in a variety of forms, including the one illustrated on page … through the *Descriptive Statistics* menu.

R code and output for one-way ANOVA with multiple comparisons

```
> bp <- read.csv("BloodPressure.csv")
> attach(bp)
> anova(lm(systolic~treatment))
Analysis of Variance Table

Response: systolic
          Df  Sum Sq Mean Sq F value    Pr(>F)
treatment  3 1913.59  637.86  24.639 5.215e-08 ***
Residuals 28  724.88   25.89
---
Signif. codes:  0 '***' 0.001 '**' 0.01 '*' 0.05 '.' 0.1 ' ' 1
> aov_res <- aov(systolic~treatment)
> # Tukey's HSD
```

```
> TukeyHSD(aov_res, "treatment", ordered = TRUE)
  Tukey multiple comparisons of means
   95% family-wise confidence level
   factor levels have been ordered


Fit: aov(formula = systolic ~ treatment)


$treatment
                        diff         lwr     upr     p adj
pills white-pills blue  8.875   1.9290003 15.821 0.0083280
no pills-pills blue    16.125   9.1790003 23.071 0.0000043
pills red-pills blue   20.375  13.4290003 27.321 0.0000001
no pills-pills white    7.250   0.3040003 14.196 0.0383062
pills red-pills white  11.500   4.5540003 18.446 0.0005651
pills red-no pills      4.250  -2.6959997 11.196 0.3575771


> plot(TukeyHSD(aov_res, "treatment"))
> # Dunnett's tests, versus control (no pills group) only
> install.packages("multcomp")
> library(multcomp)
Loading required package: mvtnorm
Loading required package: survival
Loading required package: splines
Loading required package: TH.data
> test.dunnett=glht(aov_res,linfct=mcp(treatment="Dunnett"))
> confint(test.dunnett)


        Simultaneous Confidence Intervals


Multiple Comparisons of Means: Dunnett Contrasts


Fit: aov(formula = systolic ~ treatment)


Quantile = 2.4822
95% family-wise confidence level


Linear Hypotheses:
                        Estimate lwr       upr
pills blue - no pills == 0  -16.1250 -22.4398  -9.8102
pills red - no pills == 0     4.2500  -2.0648  10.5648
pills white - no pills == 0  -7.2500 -13.5648  -0.9352


> plot(test.dunnett)
> #Fisher's LSD for a priori comparisons
> pairwise.t.test(systolic,treatment,p.adj="none")
```

```
        Pairwise comparisons using t tests with pooled SD

data:  systolic and treatment

          no pills pills blue pills red
pills blue  7.4e-07  -          -
pills red   0.1059   1.0e-08    -
pills white 0.0081   0.0016     0.0001


P value adjustment method: none
> # Bonferroni
> pairwise.t.test(systolic,treatment,p.adj="bonf")

        Pairwise comparisons using t tests with pooled SD

data:  systolic and treatment

          no pills pills blue pills red
pills blue  4.4e-06  -          -
pills red   0.63569  6.1e-08    -
pills white 0.04870  0.00974    0.00062
P value adjustment method: bonferroni
> detach(bp)
```

SPSS for one-way ANOVA with multiple comparisons

```
UNIANOVA systolic BY treatment
  /METHOD=SSTYPE(3)
  /INTERCEPT=INCLUDE
  /POSTHOC=treatment(TUKEY LSD BONFERRONI DUNNETT(1))
  /EMMEANS=TABLES(treatment)
  /PRINT=ETASQ HOMOGENEITY
  /CRITERIA=ALPHA(.05)
  /DESIGN=treatment.
```

Levene's Test of Equality of Error Variances[a]			
Dependent Variable: systolic			
F	df1	df2	Sig.
.556	3	28	.649

Tests the null hypothesis that the error variance of the dependent variable is equal across groups.

[a] Design: Intercept + treatment

Tests of Between-Subjects Effects

Dependent Variable: systolic

Source	Type III Sum of Squares	Df	Mean Square	F	Sig.	Partial Eta Squared
Corrected model	1913.594[a]	3	637.865	24.639	.000	.725
Intercept	615772.531	1	615772.531	23785.661	.000	.999
Treatment	1913.594	3	637.865	24.639	.000	.725
Error	724.875	28	25.888			
Total	618411.000	32				
Corrected total	2638.469	31				

[a] R Squared = .725 (Adjusted R Squared = .696)

treatment

Dependent Variable: systolic

Treatment	Mean	Std. Error	95% Confidence Interval Lower Bound	95% Confidence Interval Upper Bound
no pills	143.500	1.799	139.815	147.185
white pills	136.250	1.799	132.565	139.935
red pills	147.750	1.799	144.065	151.435
blue pills	127.375	1.799	123.690	131.060

Multiple Comparisons

Dependent Variable: systolic

	(I) treatment	(J) treatment	Mean difference (I-J)	Std. Error	Sig.	95% Confidence Interval Lower Bound	95% Confidence Interval Upper Bound
Tukey HSD	no pills	white pills	7.25*	2.544	.038	.30	14.20
		red pills	−4.25	2.544	.358	−11.20	2.70
		blue pills	16.13*	2.544	.000	9.18	23.07
	white pills	no pills	−7.25*	2.544	.038	−14.20	−.30
		red pills	−11.50*	2.544	.001	−18.45	−4.55
		blue pills	8.88*	2.544	.008	1.93	15.82
	red pills	no pills	4.25	2.544	.358	−2.70	11.20
		white pills	11.50*	2.544	.001	4.55	18.45
		blue pills	20.38*	2.544	.000	13.43	27.32
	blue pills	no pills	−16.13*	2.544	.000	−23.07	−9.18
		white pills	−8.88*	2.544	.008	−15.82	−1.93
		red pills	−20.38*	2.544	.000	−27.32	−13.43

Multiple Comparisons							
Dependent Variable: systolic							
			Mean			95% Confidence Interval	
	(I) treatment	(J) treatment	difference (I-J)	Std. Error	Sig.	Lower Bound	Upper Bound
LSD	no pills	white pills	7.25*	2.544	.008	2.04	12.46
		red pills	−4.25	2.544	.106	−9.46	.96
		blue pills	16.13*	2.544	.000	10.91	21.34
	white pills	no pills	−7.25*	2.544	.008	−12.46	−2.04
		red pills	−11.50*	2.544	.000	−16.71	−6.29
		blue pills	8.88*	2.544	.002	3.66	14.09
	red pills	no pills	4.25	2.544	.106	−.96	9.46
		white pills	11.50*	2.544	.000	6.29	16.71
		blue pills	20.38*	2.544	.000	15.16	25.59
	blue pills	no pills	−16.13*	2.544	.000	−21.34	−10.91
		white pills	−8.88*	2.544	.002	−14.09	−3.66
		red pills	−20.38*	2.544	.000	−25.59	−15.16
Bonferroni	no pills	white pills	7.25*	2.544	.049	.03	14.47
		red pills	−4.25	2.544	.636	−11.47	2.97
		blue pills	16.13*	2.544	.000	8.90	23.35
	white pills	no pills	−7.25*	2.544	.049	−14.47	−.03
		red pills	−11.50*	2.544	.001	−18.72	−4.28
		blue pills	8.88*	2.544	.010	1.65	16.10
	red pills	no pills	4.25	2.544	.636	−2.97	11.47
		white pills	11.50*	2.544	.001	4.28	18.72
		blue pills	20.38*	2.544	.000	13.15	27.60
	blue pills	no pills	−16.13*	2.544	.000	−23.35	−8.90
		white pills	−8.88*	2.544	.010	−16.10	−1.65
		red pills	−20.38*	2.544	.000	−27.60	−13.15
Dunnett t (2-sided)[b]	white pills	no pills	−7.25*	2.544	.022	−13.57	−.93
	red pills	no pills	4.25	2.544	.245	−2.07	10.57
	blue pills	no pills	−16.13*	2.544	.000	−22.44	−9.81

Based on observed means.
The error term is Mean Square(Error) = 25.888.
* The mean difference is significant at the .05 level.
[b] Dunnett *t*-tests treat one group as a control, and compare all other groups against it.

systolic					
				Subset	
	Treatment	N	1	2	3
Tukey HSD[a,b]	blue pills	8	127.38		
	white pills	8		136.25	
	no pills	8			143.50
	red pills	8			147.75
	Sig.		1.000	1.000	.358

Means for groups in homogeneous subsets are displayed.
Based on observed means.
The error term is Mean Square(Error) = 25.888.
[a] Uses Harmonic Mean Sample Size = 8.000.
[b] Alpha = .05.

R for two-way ANOVA – Hierarchical Nested

Nested ANOVA on Superoxide activity and Age, with Cage entered as a random factor

```
> # Nested ANOVA of superoxide activity according to
factors Age (fixed) and Cage (random)
> SO <- read.csv("Superoxide.csv")
> attach(SO)
> summary(aov(SO_activity ~ Age + Error(Cage:Age)))


Error: Cage:Age
          Df Sum Sq Mean Sq F value Pr(>F)
Age        3   1030   343.4   2.898 0.0789 .
Residuals 12   1422   118.5
---
Signif. codes:  0 '***' 0.001 '**' 0.01 '*' 0.05 '.' 0.1 ' ' 1


Error: Within
          Df Sum Sq Mean Sq F value Pr(>F)
Residuals 32   2990   93.43
Warning message:
In aov(SO_activity ~ Age + Error(Cage:Age)) : Error() model is singular
> # ANOVA without Cage
> summary(aov(SO_activity~Age))
          Df Sum Sq Mean Sq F value Pr(>F)
Age        3   1030   343.4   3.425 0.0251 *
Residuals 44   4412   100.3
---
Signif. codes:  0 '***' 0.001 '**' 0.01 '*' 0.05 '.' 0.1 ' ' 1
> detach(SO)
```

SPSS for two-way ANOVA – Hierarchical Nested

```
UNIANOVA SO_activity BY Age Cage
  /RANDOM=Cage
  /METHOD=SSTYPE(3)
  /INTERCEPT=INCLUDE
  /EMMEANS=TABLES(Age)
  /PRINT=ETASQ HOMOGENEITY
  /CRITERIA=ALPHA(.05)
  /DESIGN=Age Age*Cage.
```

Levene's Test of Equality of Error Variances[a]			
Dependent Variable: Superoxide activity			
F	df1	df2	Sig.
2.302	15	32	.023

Tests the null hypothesis that the error variance of the dependent variable is equal across groups.

[a] Design: Intercept + Age + Cage + Age * Cage

Tests of Between-Subjects Effects							
Dependent Variable: Superoxide activity							
Source		Type III Sum of Squares	df	Mean Square	F	Sig.	Partial Eta Squared
Intercept	Hypothesis	185876.032	1	185876.032	1568.641	.000	.992
	Error	1421.940	12	118.495[a]			
Age	Hypothesis	1030.179	3	343.393	2.898	.079	.420
	Error	1421.940	12	118.495[a]			
Age * Cage	Hypothesis	1421.940	12	118.495	1.268	.284	.322
	Error	2989.802	32	93.431[b]			

[a] MS(Age * Cage)
[b] MS(Error)

Age				
Dependent Variable: Superoxide activity				
			95% Confidence Interval	
Age	Mean	Std. Error	Lower Bound	Upper Bound
4 weeks	54.850	2.790	49.166	60.534
10 weeks	62.171	2.790	56.487	67.855
26 weeks	64.578	2.790	58.895	70.262
78 weeks	67.316	2.790	61.632	73.000

For comparison, do a normal one-way ANOVA of Superoxide activity and Age

```
DATASET ACTIVATE DataSet1.
GET
  FILE='C:
SuperOxide.sav'.
DATASET NAME DataSet4 WINDOW=FRONT.
ONEWAY SO_activity BY Age
/MISSING ANALYSIS.
```

ANOVA					
Superoxide activity					
	Sum of Squares	df	Mean Square	F	Sig.
Between Groups	1030.179	3	343.393	3.425	.025
Within Groups	4411.742	44	100.267		
Total	5441.921	47			

R for two-way ANOVA

Two-way ANOVA on systolic blood pressure data with two fixed factors: treatment and duration

```
> pills <- read.csv("Pills_2-way_ANOVA.csv")
> attach(pills)
The following objects are masked from pills (position 3):

 duration, systolic, treatment
> summary(aov(systolic~treatment*duration))
                 Df Sum Sq Mean Sq F value   Pr(>F)
treatment         3 1913.6   637.9  23.744 2.32e-07 ***
duration          1    3.8     3.8   0.141    0.711
treatment:duration 3   76.3    25.4   0.947    0.433
Residuals        24  644.8    26.9
---
Signif. codes:  0 '***' 0.001 '**' 0.01 '*' 0.05 '.' 0.1 ' ' 1
> # Tukey's HSD
> TukeyHSD(aov_res, "treatment", ordered = TRUE)
  Tukey multiple comparisons of means
    95% family-wise confidence level
    factor levels have been ordered

Fit: aov(formula = systolic ~ treatment * duration)

$treatment
```

```
                         diff          lwr      upr      p adj
white pills-blue pills  8.875   1.7259197 16.02408 0.0111236
no pills-blue pills     16.125  8.9759197 23.27408 0.0000112
red pills-blue pills    20.375 13.2259197 27.52408 0.0000002
no pills-white pills     7.250  0.1009197 14.39908 0.0459998
red pills-white pills   11.500  4.3509197 18.64908 0.0009361
red pills-no pills       4.250 -2.8990803 11.39908 0.3762915

> plot(TukeyHSD(aov_res, "treatment"))

> detach(pills)
```

SPSS for two-way ANOVA

```
UNIANOVA systolic BY treatment duration
  /METHOD=SSTYPE(3)
  /INTERCEPT=INCLUDE
  /POSTHOC=treatment(TUKEY)
  /EMMEANS=TABLES(treatment)
  /EMMEANS=TABLES(duration)
  /EMMEANS=TABLES(treatment*duration)
  /PRINT=ETASQ HOMOGENEITY
  /CRITERIA=ALPHA(.05)
  /DESIGN=treatment duration treatment*duration.
```

Levene's Test of Equality of Error Variances[a]			
Dependent variable: systolic			
F	df1	df2	Sig.
.373	7	24	.909

Tests the null hypothesis that the error variance of the dependent variable is equal across groups.
[a] Design: Intercept + treatment + duration + treatment * duration

Tests of Between-Subjects Effects						
Dependent Variable: systolic						
Source	Type III Sum of Squares	df	Mean Square	F	Sig.	Partial Eta Squared
Corrected Model	1993.719[a]	7	284.817	10.602	.000	.756
Intercept	615772.531	1	615772.531	22921.351	.000	.999
Treatment	1913.594	3	637.865	23.744	.000	.748
Duration	3.781	1	3.781	.141	.711	.006
treatment * duration	76.344	3	25.448	.947	.433	.106
Error	644.750	24	26.865			
Total	618411.000	32				
Corrected Total	2638.469	31				

[a] R Squared = .756 (Adjusted R Squared = .684)

1. Treatment				
Dependent Variable: systolic				
			95% Confidence Interval	
Treatment	Mean	Std. Error	Lower Bound	Upper Bound
no pills	143.500	1.833	139.718	147.282
white pills	136.250	1.833	132.468	140.032
red pills	147.750	1.833	143.968	151.532
blue pills	127.375	1.833	123.593	131.157

2. duration				
Dependent Variable: systolic				
			95% Confidence Interval	
Duration	Mean	Std. Error	Lower Bound	Upper Bound
2	139.063	1.296	136.388	141.737
4	138.375	1.296	135.701	141.049

3. treatment * duration					
Dependent variable: systolic					
				95% Confidence Interval	
Treatment	duration	Mean	Std. Error	Lower Bound	Upper Bound
no pills	2	142.750	2.592	137.401	148.099
	4	144.250	2.592	138.901	149.599
white pills	2	139.250	2.592	133.901	144.599
	4	133.250	2.592	127.901	138.599
red pills	2	147.500	2.592	142.151	152.849
	4	148.000	2.592	142.651	153.349
blue pills	2	126.750	2.592	121.401	132.099
	4	128.000	2.592	122.651	133.349

Multiple Comparisons						
Dependent Variable: systolic Tukey HSD						
(I) treatment	(J) treatment	Mean difference (I-J)	Std. Error	Sig.	95% Confidence Interval	
					Lower Bound	Upper Bound
no pills	white pills	7.2500*	2.59155	.046	.1009	14.3991
	red pills	−4.2500	2.59155	.376	−11.3991	2.8991
	blue pills	16.1250*	2.59155	.000	8.9759	23.2741
white pills	no pills	−7.2500*	2.59155	.046	−14.3991	−.1009
	red pills	−11.5000*	2.59155	.001	−18.6491	−4.3509
	blue pills	8.8750*	2.59155	.011	1.7259	16.0241
red pills	no pills	4.2500	2.59155	.376	−2.8991	11.3991
	white pills	11.5000*	2.59155	.001	4.3509	18.6491
	blue pills	20.3750*	2.59155	.000	13.2259	27.5241
blue pills	no pills	−16.1250*	2.59155	.000	−23.2741	−8.9759
	white pills	−8.8750*	2.59155	.011	−16.0241	−1.7259
	red pills	−20.3750*	2.59155	.000	−27.5241	−13.2259

Based on observed means.
The error term is mean square(error) = 26.865.
*. The mean difference is significant at the .05 level.

systolic				
Tukey HSD				
		Subset		
Treatment	N	1	2	3
blue pills	8	127.3750		
white pills	8		136.2500	
no pills	8			143.5000
red pills	8			147.7500
Sig.		1.000	1.000	.376

Means for groups in homogeneous subsets are displayed.
Based on observed means.
The error term is mean square(error) = 26.865.
[a] uses harmonic mean sample size = 8.000.
[b] Alpha = .05.

R for two-way ANOVA without then with blocks

```
> mig <- read.csv("Migration.csv")
> attach(mig)
# no blocks first
> summary(aov(distance~treatment))
```

```
         Df Sum Sq Mean Sq F value Pr(>F)
treatment   3 125380   41793   2.671 0.0826 .
Residuals  16 250400   15650
---
Signif. codes:  0 '***' 0.001 '**' 0.01 '*' 0.05 '.' 0.1 ' ' 1
# Then with blocking
> summary(aov(distance ~ treatment + blocks + Error(blocks:treatment)))

Error: blocks:treatment
          Df Sum Sq Mean Sq F value   Pr(>F)
treatment  3 125380   41793   10.61 0.001080 **
blocks     4 203130   50782   12.89 0.000266 ***
Residuals 12  47270    3939
---
Signif. codes:  0 '***' 0.001 '**' 0.01 '*' 0.05 '.' 0.1 ' ' 1
Warning message:
In aov(distance ~ treatment + blocks + Error(blocks:treatment)) :
  Error() model is singular
> detach(mig)
```

SPSS for two-way ANOVA without then with blocks

```
ONEWAY distance BY treatment
  /MISSING ANALYSIS.
```

ANOVA

distance

	Sum of Squares	df	Mean Square	F	Sig.
Between Groups	125380.000	3	41793.333	2.671	.083
Within Groups	250400.000	16	15650.000		
Total	375780.000	19			

```
UNIANOVA distance BY treatment blocks
  /RANDOM=blocks
  /METHOD=SSTYPE(3)
  /INTERCEPT=INCLUDE
  /EMMEANS=TABLES(treatment)
  /PRINT=ETASQ
  /CRITERIA=ALPHA(.05)
  /DESIGN=treatment blocks treatment*blocks.
```

Tests of Between-Subjects Effects

Dependent Variable: distance

Source		Type III Sum of Squares	df	Mean Square	F	Sig.	Partial Eta Squared
Intercept	Hypothesis	4436820.000	1	4436820.000	87.369	.001	.956
	Error	203130.000	4	50782.500[a]			
treatment	Hypothesis	125380.000	3	41793.333	10.610	.001	.726
	Error	47270.000	12	3939.167[b]			
blocks	Hypothesis	203130.000	4	50782.500	12.892	.000	.811
	Error	47270.000	12	3939.167[b]			
treatment * blocks	Hypothesis	47270.000	12	3939.167	.	.	1.000
	Error	.000	0	.[c]			

[a] MS(blocks)
[b] MS(treatment * blocks)
[c] MS(error)

treatment

Dependent Variable: distance

treatment	Mean	Std. Error	95% Confidence Interval	
			Lower Bound	Upper Bound
Culture medium	366.000	.	.	.
Compound A	442.000	.	.	.
Compound B	492.000	.	.	.
Compound C	584.000	.	.	.

Chapter 7
R for regression of weight on height in swimmers

```
> hw <- read.csv("HeightWeight.csv")
> attach(hw)
> summary(lm(Weight ~ Height))

Call:
lm(formula = Weight ~ Height)
Residuals:
    Min      1Q   Median      3Q     Max
-11.0286  -2.6361  -0.5071  2.4929  17.0574
```

```
Coefficients:
            Estimate Std. Error t value Pr(>|t|)
(Intercept) -68.29881    5.60614  -12.18   <2e-16 ***
Height        0.76075    0.03247   23.43   <2e-16 ***
---
Signif. codes:  0 '***' 0.001 '**' 0.01 '*' 0.05 '.' 0.1 ' ' 1

Residual standard error: 4.303 on 358 degrees of freedom
Multiple R-squared:  0.6053,      Adjusted R-squared:  0.6041
F-statistic: 548.9 on 1 and 358 DF,  p-value: < 2.2e-16


> detach(hw)
```

SPSS for regression of weight on height in swimmers

```
REGRESSION
  /MISSING LISTWISE
  /STATISTICS COEFF OUTS R ANOVA
  /CRITERIA=PIN(.05) POUT(.10)
  /NOORIGIN
  /DEPENDENT weight
  /METHOD=ENTER height
  /SCATTERPLOT=(*ZRESID ,*ZPRED)
  /RESIDUALS HISTOGRAM(ZRESID) NORMPROB(ZRESID).
```

Model Summary[b]				
Model	R	R Square	Adjusted R Square	Std. Error of the Estimate
1	.778[a]	.605	.604	4.303

[a] Predictors: (Constant), height
[b] Dependent Variable: weight

ANOVA[a]						
Model		Sum of Squares	df	Mean Square	F	Sig.
1	Regression	10163.796	1	10163.796	548.909	.000[b]
	Residual	6628.860	358	18.516		
	Total	16792.656	359			

[a] Dependent Variable: weight
[b] Predictors: (Constant), height

Coefficients[a]						
		Unstandardised coefficients		Standardised coefficients		
Model		B	Std. Error	Beta	t	Sig.
1	(Constant)	−68.299	5.606		−12.183	.000
	height	.761	.032	.778	23.429	.000

[a] Dependent Variable: weight

Residuals Statistics[a]					
	Minimum	Maximum	Mean	Std. Deviation	N
Predicted Value	50.38	75.48	62.94	5.321	360
Residual	−11.029	17.057	.000	4.297	360
Std. Predicted Value	−2.361	2.358	.000	1.000	360
Std. Residual	−2.563	3.964	.000	.999	360

[a] Dependent Variable: weight

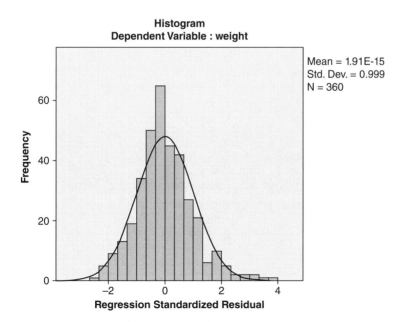

Histogram
Dependent Variable : weight

Mean = 1.91E-15
Std. Dev. = 0.999
N = 360

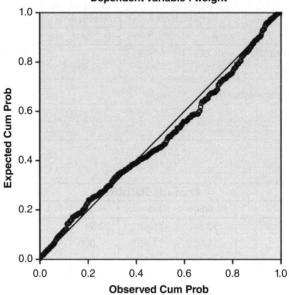

Normal P-P Plot of Regression Standardized Residual
Dependent Variable : weight

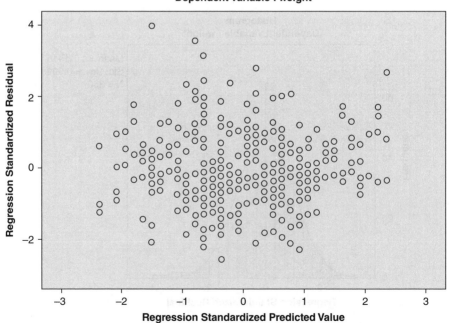

Scatterplot
Dependent Variable : weight

R for correlation between BMI and birth weight

```
> bmi <- read.csv("BMI_BirthWeight.csv")
> attach(bmi)
> cor.test(BMI,b_weight)


        Pearson's product-moment correlation

data:  BMI and b_weight
t = 14.441, df = 98, p-value < 2.2e-16
alternative hypothesis: true correlation is not equal to 0
95 percent confidence interval:
 0.7498732 0.8788456
sample estimates:
      cor
0.8248066


> detach(bmi)
```

SPSS for correlation between BMI and birth weight

```
CORRELATIONS
  /VARIABLES=BMI Birth_Weight
  /PRINT=TWOTAIL NOSIG
  /MISSING=PAIRWISE.
```

Correlations			BMI	Birth_Weight
BMI	Pearson Correlation		1	.825**
	Sig. (2-tailed)			.000
	N		100	100
Birth_Weight	Pearson Correlation		.825**	1
	Sig. (2-tailed)		.000	
	N		100	100

**Correlation is significant at the 0.01 level (2-tailed).

Minitab output for multiple regression

Regression Analysis: Heart rate versus Exercise Intensity, Activity, Type of Athlete

```
The regression equation is
Heart rate = - 45.1 + 3.27 Exercise intensity + 9.06 Activity
         + 45.6 Type of athlete
```

Predictor	Coef	SE Coef	T	P	VIF
Constant	-45.05	14.96	-3.01	0.009	
Exercise	3.271	1.436	2.28	0.039	7.4
Activity	9.061	3.328	2.72	0.017	4.7
Type of athlete	45.572	5.822	7.83	0.000	4.8

S = 5.320   R-Sq = 95.7%   R-Sq(adj) = 94.7%

Analysis of Variance

Source	DF	SS	MS	F	P
Regression	3	8767.7	2922.6	103.25	0.000
Residual Error	14	396.3	28.3		
Total	17	9164.0			

Regression Analysis: Heart rate versus Exercise intensity, Type of athlete

The regression equation is
Heart rate = - 76.1 + 6.73 Exercise intensity + 58.5 Type of athlete

Predictor	Coef	SE Coef	T	P	VIF
Constant	-76.07	11.58	-6.57	0.000	
Exercise	6.7305	0.8003	8.41	0.000	1.6
Type of athlete	58.502	4.024	14.54	0.000	1.6

S = 6.357   R-Sq = 93.4%   R-Sq(adj) = 92.5%

Analysis of Variance

Source	DF	SS	MS	F	P
Regression	2	8557.9	4279.0	105.90	0.000
Residual Error	15	606.1	40.4		
Total	17	9164.0			

R for multiple regression

```
> install.packages("car")        #need car for vif function
Installing package(s)    into    'C:/....../R/win-library/2.15'
(as 'lib' is unspecified)
trying                                                          URL
'http://www.stats.bris.ac.uk/R/bin/windows/contrib/2.15/car_2.0-19.zip'
Content type 'application/zip' length 1326306 bytes (1.3 Mb)
opened URL
downloaded 1.3 Mb


package 'car' successfully unpacked and MD5 sums checked
```

```
The downloaded binary packages are in
    C:\......\downloaded_packages
> library(car)
>
> MRdata <- read.csv("HR2.csv",header=TRUE)
>
> mr1 <- lm(HR~Athlete+Intensity+Activity,data=MRdata)    #all  3
predictors
>
> summary(mr1)


Call:
lm(formula = HR ~ Athlete + Intensity + Activity, data = MRdata)


Residuals:
    Min      1Q  Median      3Q     Max
-7.7149 -2.8288 -0.2647  1.6977 10.9783


Coefficients:
            Estimate Std. Error t value Pr(>|t|)
(Intercept)   0.5186     9.6770   0.054   0.9580
Athlete      45.5715     5.8222   7.827 1.76e-06 ***
Intensity     3.2714     1.4363   2.278   0.0390 *
Activity      9.0608     3.3279   2.723   0.0165 *
---
Signif. codes:  0 '***' 0.001 '**' 0.01 '*' 0.05 '.' 0.1 ' ' 1


Residual standard error: 5.32 on 14 degrees of freedom
Multiple R-squared: 0.9568,      Adjusted R-squared: 0.9475
F-statistic: 103.3 on 3 and 14 DF,  p-value: 8.713e-10


> vif(mr1)                              #variance inflation
factors
  Athlete Intensity  Activity
 4.790489  7.369056  4.695382
>
> mr2 <- lm(HR~Athlete+Intensity,data=MRdata)  #dropping type of
activity
> summary(mr2)
Call:
lm(formula = HR ~ Athlete + Intensity, data = MRdata)


Residuals:
   Min     1Q Median     3Q    Max
-8.195 -4.667 -1.465  6.410  8.535
```

```
Coefficients:
            Estimate Std. Error t value Pr(>|t|)
(Intercept) -17.5711    8.4060   -2.09    0.054 .
Athlete      58.5024    4.0238   14.54 3.01e-10 ***
Intensity     6.7305    0.8003    8.41 4.63e-07 ***
---
Signif. codes:  0 '***' 0.001 '**' 0.01 '*' 0.05 '.' 0.1 ' ' 1

Residual standard error: 6.357 on 15 degrees of freedom
Multiple R-squared: 0.9339,      Adjusted R-squared: 0.925
F-statistic: 105.9 on 2 and 15 DF,  p-value: 1.424e-09


> vif(mr2)
Athlete Intensity
1.602818  1.602818
>
> #setting up dummy coded variables for types of activity vs walking
> sprint <- c(0,0,0,0,0,0,0,0,0,0,1,1,1,0,0,0,1,1,1)
> jog <- c(0,0,0,0,1,1,1,1,1,0,0,0,0,0,1,0,0,0)
> MRdata <- data.frame(MRdata, jog, sprint)           #add to
dataframe
>
> mr3 <- lm(HR~Athlete+Intensity+jog+sprint,data=MRdata)  #using dummy
vars
> summary(mr3)


Call:
lm(formula = HR ~ Athlete + Intensity + jog + sprint, data = MRdata)

Residuals:
    Min      1Q  Median      3Q     Max
-8.4237 -2.6617 -0.8703  1.4364  9.2627


Coefficients:
            Estimate Std. Error t value Pr(>|t|)
(Intercept)    6.559     12.226   0.536   0.6007
Athlete       46.573      5.772   8.069 2.04e-06 ***
Intensity      3.798      1.473   2.579   0.0229 *
jog            4.570      4.884   0.936   0.3665
sprint        16.113      6.732   2.393   0.0325 *
---
Signif. codes:  0 '***' 0.001 '**' 0.01 '*' 0.05 '.' 0.1 ' ' 1

Residual standard error: 5.222 on 13 degrees of freedom
Multiple R-squared: 0.9613,      Adjusted R-squared: 0.9494
F-statistic: 80.75 on 4 and 13 DF,  p-value: 4.781e-09
```

```
> vif(mr3)
 Athlete Intensity       jog    sprint
4.886747  8.040161  3.498795  6.647390
```

3-dimensional plot done in SPSS for the multiple regression example

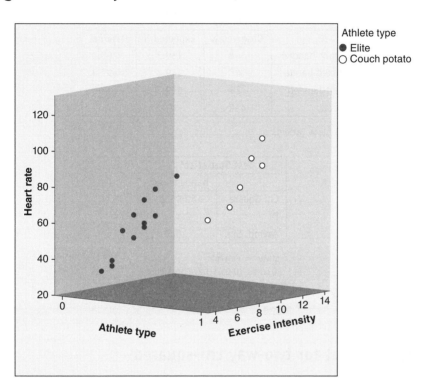

Chapter 8
Minitab output for one-way chi-squared for patients

Counts in variable: counts

```
Using category names in handedness
```

		Test		Contribution
Category	Observed	Proportion	Expected	to Chi-Sq
1	348	0.333333	139.333	312.501
2	42	0.333333	139.333	67.994
3	28	0.333333	139.333	88.960

N	DF	Chi-Sq	P-Value
418	2	469.455	0.000

SPSS for one-way chi-squared for patients

```
WEIGHT BY freq.
NPAR TESTS
  /CHISQUARE=handedness
  /EXPECTED=EQUAL
  /MISSING ANALYSIS.
```

handedness[a]			
	Observed N	Expected N	Residual
Right handed	348	139.3	208.7
Mixed handed	42	139.3	−97.3
Left handed	28	139.3	−111.3
Total	418		

[a] status = Patients

Test Statistics[a]	
	handedness
Chi-Square	469.455[b]
Df	2
Asymp. Sig.	.000

[a] status = Patients
[b] 0 cells (0.0%) have expected frequencies less than 5. The minimum expected cell frequency is 139.3.

Minitab output for two-way chi-squared

Data from Table 8.2

Tabulated statistics: gender, handedness

Using frequencies in Counts

```
Rows: Gender   Columns: Handedness

             1        2        3      All

1          221       31       23      275
        228.95    27.63    18.42   275.00
       -0.5252   0.6408   1.0669        *

2          127       11        5      143
        119.05    14.37     9.58   143.00
        0.7284  -0.8886  -1.4795        *
```

```
All        348        42        28       418
          348.00     42.00     28.00   418.00
             *          *         *         *
```

```
Cell Contents:      Count
                    Expected count
                    Standardized residual
```

Pearson Chi-Square = 5.334, DF = 2, P-Value = 0.069
Likelihood Ratio Chi-Square = 5.764, DF = 2, P-Value = 0.056

SPSS for two-way chi-squared

```
CROSSTABS
  /TABLES=gender BY handedness
  /FORMAT=AVALUE TABLES
  /STATISTICS=CHISQ
  /CELLS=COUNT
  /COUNT ROUND CELL.
```

gender * handedness Crosstabulation[a]					
Count					
		handedness			Total
		Right handed	Mixed handed	Left handed	
gender	Men	221	31	23	275
	Women	127	11	5	143
Total		348	42	28	418

[a] status = Patients

Chi-Square tests[a]			
	Value	df	Asymp. Sig. (2-sided)
Pearson Chi-Square	5.334[b]	2	.069
Likelihood Ratio	5.764	2	.056
Linear-by-Linear Association	5.308	1	.021
N of Valid Cases	418		

[a] status = Patients
[b] 0 cells (0.0%) have expected count less than 5. The minimum expected count is 9.58.

Minitab output for non-patients' data

Data from Table 8.3

Tabulated statistics: gender, handedness

Using frequencies in Counts

```
Rows: Gender    Columns: Handedness

            1        2        3      All

1         109       13       13      135
       114.84    11.72     8.44   135.00
      -0.5453   0.3743   1.5707        *

2         136       12        5      153
       130.16    13.28     9.56   153.00
       0.5122  -0.3516  -1.4754        *

All       245       25       18      288
       245.00    25.00    18.00   288.00
            *        *        *        *

Cell Contents:       Count
                     Expected count
                     Standardized residual

Pearson Chi-Square = 5.467, DF = 2, P-Value = 0.065
Likelihood Ratio Chi-Square = 5.579, DF = 2, P-Value = 0.061
```

SPSS for non-patients' data

gender * handedness Crosstabulation[a]					
Count					
		handedness			
		Right handed	Mixed handed	Left handed	Total
gender	Men	109	13	13	135
	Women	136	12	5	153
Total		245	25	18	288

[a] status = Non-patients

Chi-Square Tests[a]			
	Value	df	Asymp. Sig. (2-sided)
Pearson Chi-Square	5.467[b]	2	.065
Likelihood Ratio	5.579	2	.061
Linear-by-Linear Association	5.154	1	.023
N of Valid Cases	288		

[a] status = Non-patients
[b] 0 cells (0.0%) have expected count less than 5. The minimum expected count is 8.44.

Minitab output for patients versus non-patients, two-way chi-squared

Tabulated statistics: status, handedness

Using frequencies in Counts

```
Rows: Status    Columns: Handedness

              1         2         3       All

1           245        25        18       288
         241.90     27.33     18.76    288.00
         0.1991   -0.4460   -0.1766         *

2           348        42        28       418
         351.10     39.67     27.24    418.00
        -0.1652    0.3702    0.1466         *

All         593        67        46       706
         593.00     67.00     46.00    706.00
              *         *         *         *


Cell Contents:      Count
                    Expected count
                    Standardized residual

Pearson Chi-Square = 0.456, DF = 2, P-Value = 0.796
Likelihood Ratio Chi-Square = 0.459, DF = 2, P-Value = 0.795
```

Data Display

```
Row   Status   Handedness    Counts
  1        1            1       245
  2        1            2        25
  3        1            3        18
  4        2            1       348
  5        2            2        42
  6        2            3        28
```

SPSS for patients versus non-patients, two-way chi-squared

```
CROSSTABS
  /TABLES=status BY handedness
  /FORMAT=AVALUE TABLES
  /STATISTICS=CHISQ
  /CELLS=COUNT
  /COUNT ROUND CELL.
```

status * handedness Crosstabulation					
Count					
		Handedness			Total
		Right handed	Mixed handed	Left handed	
Status	Non-patients	245	25	18	288
	Patients	348	42	28	418
Total		593	67	46	706

Chi-Square Tests			
	Value	df	Asymp. Sig. (2-sided)
Pearson Chi-Square	.456[a]	2	.796
Likelihood Ratio	.459	2	.795
Linear-by-Linear Association	.286	1	.593
N of Valid Cases	706		

[a] 0 cells (0.0%) have expected count less than 5. The minimum expected count is 18.76.

Minitab output for Gender and handedness (patients and non-patients combined), two-way chi-squared test

Data in Table 8.4

Gender and handedness, patients and non-patients combined

```
Using frequencies in Counts

Rows: Gender    Columns: Handedness

              1         2         3       All

1           330        44        36       410
           344.38     38.91     26.71    410.00
          -0.7747    0.8161    1.7967        *
```

```
2           263        23        10     296
          248.62     28.09     19.29  296.00
          0.9118   -0.9605   -2.1145      *


All         593        67        46     706
          593.00     67.00     46.00  706.00
              *         *         *        *


Cell Contents:      Count
                    Expected count
                    Standardized residual


Pearson Chi-Square = 10.719, DF = 2, P-Value = 0.005
Likelihood Ratio Chi-Square = 11.391, DF = 2, P-Value = 0.003
```

SPSS for Gender and handedness (patients and non-patients combined), two-way chi-squared test

```
CROSSTABS
   /TABLES=gender BY handedness
   /FORMAT=AVALUE TABLES
   /STATISTICS=CHISQ
   /CELLS=COUNT
   /COUNT ROUND CELL.
```

gender * handedness Crosstabulation					
Count					
		handedness			
		Right handed	Mixed handed	Left handed	Total
gender	Men	330	44	36	410
	Women	263	23	10	296
Total		593	67	46	706

Chi-Square Tests			
	Value	df	Asymp. Sig. (2-sided)
Pearson Chi-Square	10.719[a]	2	.005
Likelihood Ratio	11.391	2	.003
Linear-by-Linear Association	10.672	1	.001
N of Valid Cases	706		

[a] 0 cells (0.0%) have expected count less than 5. The minimum expected count is 19.29.

Minitab output for Fisher's exact test
Tabulated statistics: gender, colour

```
Using frequencies in frequency

Rows: Gender   Columns: Colour

        1  2  All

1       7  0   7
2       4  4   8
All    11  4  15

Cell Contents:     Count

Fisher's exact test: P-Value =  0.0769231
```

SPSS for Fisher's exact test

```
WEIGHT BY freq.
CROSSTABS
  /TABLES=gender BY colour
  /FORMAT=AVALUE TABLES
  /STATISTICS=CHISQ
  /CELLS=COUNT EXPECTED
  /COUNT ROUND CELL.
```

gender * colour Crosstabulation					
			colour		Total
			Blue	Pink	
Gender	Boys	Count	7	0	7
		Expected count	5.1	1.9	7.0
	Girls	Count	4	4	8
		Expected count	5.9	2.1	8.0
Total		Count	11	4	15
		Expected count	11.0	4.0	15.0

Chi-Square Tests					
	Value	Df	Asymp. Sig. (2-sided)	Exact Sig. (2-sided)	Exact Sig. (1-sided)
Pearson Chi-Square	4.773[a]	1	.029		
Continuity Correction[b]	2.558	1	.110		
Likelihood Ratio	6.307	1	.012		
Fisher's Exact Test				.077	.051
Linear-by-Linear Association	4.455	1	.035		
N of Valid Cases	15				

[a] 2 cells (50.0%) have expected count less than 5. The minimum expected count is 1.87.
[b] Computed only for a 2 x 2 table

SPSS for odds ratio, effects of folic acid on neural tube defects

```
WEIGHT BY freq.
CROSSTABS
  /TABLES=treatment BY outcome
  /FORMAT=AVALUE TABLES
  /STATISTICS=CHISQ RISK
  /CELLS=COUNT
  /COUNT ROUND CELL.
```

treatment * Neural tube defects Crosstabulation				
Count				
		Neural tube defects		Total
		Yes	No	
treatment	Folic acid	6	587	593
	None	21	581	602
Total		27	1168	1195

Chi-Square Tests					
	Value	Df	Asymp. Sig. (2-sided)	Exact Sig. (2-sided)	Exact Sig. (1-sided)
Pearson Chi-Square	8.297[a]	1	.004		
Continuity Correction[b]	7.213	1	.007		
Likelihood Ratio	8.789	1	.003		
Fisher's Exact Test				.005	.003
Linear-by-Linear Association	8.290	1	.004		
N of Valid Cases	1195				

[a] 0 cells (0.0%) have expected count less than 5. The minimum expected count is 13.40.
[b] Computed only for a 2 x 2 table

Risk Estimate			
		95% Confidence Interval	
	Value	Lower	Upper
Odds Ratio for treatment (Folic acid/None)	.283	.113	.706
For cohort Neural tube defects = Yes	.290	.118	.714
For cohort Neural tube defects = No	1.026	1.008	1.043
N of Valid Cases	1195		

Chapter 9

Minitab output for binomial test

Sign test for median: WBC

Sign test of median = 7.000 versus not = 7.000

```
        N  Below  Equal  Above       P  Median
WBC  10      9      0      1  0.0215   4.500
```

SPSS for binomial test

```
NPAR TESTS
  /BINOMIAL (0.50)=WBC (7)
  /MISSING ANALYSIS.
```

		Binomial Test				
	Category	N	Observed Prop.	Test Prop.	Exact Sig. (2-tailed)	
WBC	Group 1	<= 7	9	.90	.50	.021
	Group 2	>7	1	.10		
	Total		10	1.00		

Minitab output for Mann–Whitney test

Data from Table 9.4

```
          N  Median
Crocus    9    0.00
Placebo  10    3.00
```

Point estimate for ETA1-ETA2 is -2.00
95.5 Percent CI for ETA1-ETA2 is (-3.99,-0.99)
W = 62.0
Test of ETA1 = ETA2 vs ETA1 not = ETA2 is significant at 0.0247
The test is significant at 0.0225 (adjusted for ties)

SPSS for Mann–Whitney test

```
NPAR TESTS
  /M-W= spots BY treatment(1 2)
  /MISSING ANALYSIS.
```

Ranks				
	treatment	N	Mean Rank	Sum of Ranks
Spots	Crocus	9	6.89	62.00
	Placebo	10	12.80	128.00
	Total	19		

Test Statistics[a]	
	spots
Mann-Whitney U	17.000
Wilcoxon W	62.000
Z	−2.323
Asymp. Sig. (2-tailed)	.020
Exact Sig. [2*(1-tailed Sig.)]	.022[b]

[a] Grouping variable: treatment
[b] Not corrected for ties.

Minitab output for Wilcoxon signed rank test for paired samples (actually a one-sample Wilcoxon on the column of differences between the two treatments)

Wilcoxon signed rank test: differences

```
Test of median = 0.000000 versus median not = 0.000000


                  N for   Wilcoxon            Estimated
             N    Test    Statistic      P     Median
Differences  8     8          7.0    0.141    -5.000
```

SPSS for Wilcoxon signed rank test for paired samples

```
NPAR TESTS
  /WILCOXON=HotWater WITH Coffee (PAIRED)
  /STATISTICS QUARTILES
  /MISSING ANALYSIS.
```

Descriptive Statistics				
		Percentiles		
	N	25th	50th (Median)	75th
HotWater	8	1.00	2.00	3.75
Coffee	8	4.50	7.50	9.75

Ranks					
		N	Mean Rank	Sum of Ranks
Coffee - HotWater	Negative Ranks	1[a]	7.00	7.00
	Positive Ranks	7[b]	4.14	29.00
	Ties	0[c]		
	Total	8		

[a] Coffee < Hotwater
[b] Coffee > Hotwater
[c] Coffee = Hotwater

Test Statistics[a]	
	Coffee - HotWater
Z	−1.556[b]
Asymp. Sig. (2-tailed)	.120

[a] Wilcoxon signed ranks test
[b] Based on negative ranks.

Minitab output for Kruskal–Wallis one-way analysis of variance

Kruskal–Wallis test: jump height versus flea species

```
Kruskal-Wallis Test on Jump Height


Flea
species   N  Median  Ave Rank      Z
cats      9  20.000     14.1    0.30
dogs      9  36.000     17.8    2.08
humans    8   5.000      8.0   -2.44
Overall  26             13.5


H = 7.01   DF = 2   P = 0.030
```

SPSS for Kruskal–Wallis one-way analysis of variance

```
DATASET ACTIVATE DataSet13.
NPAR TESTS
  /K-W=Jump BY Flea(1 3)
  /MISSING ANALYSIS.
```

Ranks			
	Species of flea	N	Mean Rank
Height in cm	Cat	9	14.11
	Dog	9	17.78
	Human	8	8.00
	Total	26	

Test Statistics[a],[b]	
	Height in cm
Chi-Square	7.031
df	2
Asymp. Sig.	.030

[a] Kruskal–Wallis test
[b] Grouping Variable: species of flea

Minitab output for Friedman's related samples test

Data from Table 9.10

Treatments 1 – hot water, 2 – decaff coffee, 3 – coffee

```
S = 6.75  DF = 2  P = 0.034

                 Est   Sum of
Treatments  N  Median   Ranks
1           8   1.833    13.0
2           8   2.000    13.0
3           8   7.167    22.0

Grand median = 3.667
```

SPSS for Friedman's related samples test

```
NPAR TESTS
  /FRIEDMAN=HotWater DecaffCoffee Coffee
  /STATISTICS QUARTILES
  /MISSING LISTWISE.
```

Descriptive Statistics				
		Percentiles		
	N	25th	50th (Median)	75th
HotWater	8	1.00	2.00	3.75
Decaffcoffee	8	.25	2.00	3.00
Coffee	8	4.50	7.50	9.75

Ranks	
	Mean rank
HotWater	1.63
Decaffcoffee	1.63
Coffee	2.75

Test Statistics[a]	
N	8
Chi-Square	6.750
Df	2
Asymp. Sig.	.034

[a] Friedman Test

Index

Starting Out in Statistics: An Introduction for Students of Human Health, Disease, and Psychology
First Edition. Patricia de Winter and Peter M. B. Cahusac.
© 2014 John Wiley & Sons, Ltd. Published 2014 by John Wiley & Sons, Ltd.
Companion Website: www.wiley.com/go/deWinter/startingstatistics